THE
ORACLE
OF OIL

THE ORACLE OF OIL

A Maverick Geologist's Quest
for a Sustainable Future

MASON INMAN

W. W. NORTON & COMPANY
Independent Publishers Since 1923
New York · London

For information about permission to reproduce selections from this book,
write to Permissions, W. W. Norton & Company, Inc.,
500 Fifth Avenue, New York, NY 10110

For information about special discounts for bulk purchases, please contact
W. W. Norton Special Sales at specialsales@wwnorton.com or 800-233-4830

Manufacturing by Quad Graphics Fairfield
Book design by Ellen Cipriano
Production manager: Louise Mattarelliano

Library of Congress Cataloging-in-Publication Data

Names: Inman, Mason.
Title: The oracle of oil : a maverick geologist's quest for a sustainable
    future / Mason Inman.
Description: New York : W.W. Norton & Company, Inc., [2016] | Includes
    bibliographical references and index.
Identifiers: LCCN 2015049966 | ISBN 9780393239683 (hardcover)
Subjects: LCSH: Hubbert, M. King (Marion King), 1903-1989. | Petroleum
    geologists—United States—Biography. | Petroleum reserves—Forecasting. |
    Hubbert peak theory.
Classification: LCC TN869.2.H83 I56 2016 | DDC 553.2/8092—dc23 LC record
available at http://lccn.loc.gov/2015049966

W. W. Norton & Company, Inc.
500 Fifth Avenue, New York, N.Y. 10110
www.wwnorton.com

W. W. Norton & Company Ltd.
Castle House, 75/76 Wells Street, London W1T 3QT

1 2 3 4 5 6 7 8 9 0

To my mom,
for her endless support

To my dad,
who would have enjoyed this book

CONTENTS

PART 4: HOUSTON, TEXAS, AND WASHINGTON, DC, 1956–1973

PART 5: WASHINGTON, DC, 1973–1989

PROLOGUE

MARION KING HUBBERT HAD BEEN preparing for this moment for months—or in some ways, for thirty years, nearly his entire adult life. On March 8, 1956, at about ten in the morning, he was sitting on the stage in a long, narrow hall, the top-floor ballroom of San Antonio's Plaza Hotel, facing an audience of some five hundred oilmen. At this meeting of the American Petroleum Institute, Hubbert would give the keynote speech. The day of his talk he had taken extra care with his attire and had tamed his bristly hair, slicking it back from his large forehead.

He had planned to make a bold prediction. He'd tell the audience that, based on an in-depth analysis he'd done of US oil resources, the nation was approaching a crisis. American oil production—the rate at which it extracted crude oil from the ground—had been soaring ever higher for a century. But according to Hubbert's calculations, this trend would soon end. Within ten to fifteen years—by the early 1970s at the latest—he expected US oil production would peak. For the whole world, he figured, this peak would come later, around the turn of the twenty-first century. After each of these peaks, production would likely dwindle, with less and less oil available year after year. The peak wouldn't mean the end of oil, but it would mark a crucial turning point, from an age of abundance to one of scarcity.

Just before he was to step up to the stout wood podium and deliver his speech, someone signaled for him to get up and leave the room. He rushed out of the hall to find an urgent call waiting from his employer,

Shell Oil. On the other end of the line was an executive assistant in the public relations office in New York. Headquarters wanted Hubbert to change his message.

"Couldn't you tone it down some?" the man pleaded. "Couldn't you take the sensational parts out?"

"Nothing sensational about it," Hubbert replied. "Just straightforward analysis."

"That part about reaching the peak of oil production in ten or fifteen years, it's just utterly ridiculous."

OIL IS SO completely woven into modern society, it can be difficult to grasp how dependent we are on it. We've generally assumed that it will continue to be plentiful for the foreseeable future. On that assumption, we've planned our cities—and often let them sprawl without much planning. Our daily commutes are fueled by oil, along with our globe-trotting vacations. Our shopping carts are filled with groceries from around the world, thanks to oil. In 2015, Americans burned about 20 million barrels a day, or nearly 7 billion barrels a year: that is, each American consumes about two and a half gallons each day.

Oil has been with us more than a century and a half, ever since an 1859 well in Pennsylvania kicked off the oil age. For the past half-century, oil has been the world's number-one fuel, consumed more voraciously than coal or natural gas and far more than all the other alternatives. Most of the oil consumed worldwide is known as conventional crude oil—the kind where you can drill a hole in the ground and pump it out. For the past decade, starting in 2006, production of plain old crude has been stuck roughly on a plateau, at around 70 million barrels a day—despite oil prices having soared to record highs. So the current director of the International Energy Agency, Fatih Birol, has declared, "The age of cheap oil is over."

Since 2006 increases in total world oil production have been entirely reliant on what the industry calls "unconventional" sources. One of the oldest of these unconventional methods—employed by the Nazis during the Second World War and still in use in South Africa—is to turn coal into liquid fuel. Later came the rise of tar sands, principally in Canada, recently surpassing 2 million barrels a day. Biofuels—such as ethanol made from corn and diesel made from soy—supply more than a million additional barrels a day.

The latest unconventional source is known as shale oil, opened up by hydraulic fracturing. This technique, also called fracking, has been around for decades. (Incidentally, in the 1950s, Hubbert was the first one to correctly explain how fracking works.) Fracking has recently seen a huge expansion. Coupled with advances in horizontal drilling, it has been applied to tens of thousands of wells in the United States, and in less than five years, it boosted the nation's production by more than 4 million barrels per day, nearly doubling it and turning around a decades-long decline. The reversal was so dramatic, many called this boom "the shale revolution."

Just since 2010 investors have poured hundreds of billions of dollars into developing unconventional oil. Yet all these alternatives combined still make up only a small fraction of the world's total oil supply, less than one-tenth. The vast majority of the world's oil still comes from conventional sources, which have been struggling just to maintain their production rates and have even started to decline slightly.

The past several years have been a bumpy ride for the world economy. After a long run of high oil prices came the "Great Recession" in 2008 and its lasting aftermath, the worst economic calamity in several decades. When oil prices crashed in 2014, the weak economy was a major factor. These economic woes and the roller coaster of oil prices may both be symptoms of a world near the crest of oil production, unable to boost production substantially at prices that society is willing or able to bear.

KING HUBBERT TRIED, throughout his career, to forecast limits to oil production. More broadly, he tried to warn the world about other such limits and to point the way toward a sustainable future.

Today experts still argue over whether Hubbert's oil forecasts were correct—an issue that, it turns out, is no simple matter to decide. It depends partly on cutting through the misrepresentations and glosses of both his critics and his supporters, to understand what he did and did not say. But more deeply, it depends on what it means for a prediction to be correct. It's one thing to predict the outcome of an election, where there's one clear winner. Foretelling the long-term future of the modern world's top fuel is a messier endeavor. To be considered correct (or at least useful), does a prediction have to nail every detail? Is it enough to be close?

When discussing the nature of predictions, Hubbert liked to tell the story of Croesus, who in the sixth century B.C. ruled the kingdom of Lydia, in modern-day Turkey. When Croesus was considering launching a war to topple the neighboring Persian Empire, he sent messengers to the Athenian temple at Delphi to consult its oracles. These women soothsayers, who sat in the temple's basement and went into a trance before delivering their forecasts, gave Croesus this prophecy: If he sent his army against Persia, he would destroy a great empire. Croesus took this as a positive sign and went ahead. The oracles were right—but the kingdom that fell was Croesus's own. All too often, Hubbert pointed out, predictions for the future are like the oracles': vague and open to varying interpretations.

There was another part to the story that Hubbert didn't tell. After losing the war, Croesus was distraught. Before taking the oracles' advice, he'd put them to a test. He'd sent messengers to several cities—in modern-day Greece, Libya, and other nearby lands. On a prearranged day and time, the messengers asked the local prophets, sages, and diviners what Croesus was doing at that moment. On that day, he chose to do something odd. He boiled a tortoise and a lamb in a bronze pot. Only the oracles at Delphi gave the correct answer. From then on, Croesus trusted them—until his downfall.

Today we have much the same attitude toward prediction and those who predict. We want oracles, people whose predictions we can trust, even if we don't understand how they arrive at their conclusions. Like Croesus, we welcome predictions that sound like good news, and we don't stop to ask questions. Later, if things go badly, we feel betrayed by our trusted seers.

Hubbert didn't mind being the bearer of bad news. But throughout most of his career, his warnings were ignored or dismissed. And even when his forecasts began to be taken seriously, his followers didn't pay much attention to his methods or his caveats. Rather than engaging with his forecasts and building on them or constructively critiquing them, most of his followers and his critics treated him like an oracle, one whose forecasts could be either believed or dismissed but whose methods were unfathomable.

But Hubbert was no oracle, of course. He was a scientist. When making predictions, he always presented his data and reasoning for anyone to examine for themselves. It was the same approach he used in all his scientific studies—a body of work that earned him a reputation as "twentieth-century geology's Renaissance man." A couple of years after his death, a 1991 report from the National Academy of Sciences summed up his legacy thus: "This outspoken maverick led the earth sciences kicking and screaming from a largely observational and descriptive style to a more quantitative, experimental, predictive science."

Despite the solid reputation Hubbert earned for his geological work, many of his colleagues looked askance at his predictions for oil. Even fewer bought into his critique of growth—both of populations and economies. Despite such skepticism, throughout his life Hubbert tried to point the way toward a radically different future. He envisioned a world without growth and with a stable world global population, stable levels of consumption, and an environment that, rather than deteriorating, was likewise stable.

Making a transition from our current ways to such a steady-state society would, Hubbert believed, require an intellectual revolution. If humanity managed it, he hoped it would lead to an equal and just society, allowing for lives of leisure and learning. It would be a cultural renaissance unlike any the world had ever experienced.

Hubbert's oil forecasts earned him a reputation as a pessimist. In fact he was a utopian.

PART 1

Chicago, 1923–1929

1

The Journey

IN THE FALL OF 1923, King Hubbert's train sped toward Chicago, passing sordid suburbs, factories, and foundries, jostling as it crossed innumerable other rail lines. As it approached the metropolis, Chicago's defining feature came into Hubbert's view: a great, thick pall of soot hanging over the city. Thick smoke poured from factory chimneys, from steamers churning along rivers, and from power plants that lit skyscrapers. The dense fumes, from burning sulfurous brown coal, corroded building facades and bridges. To combat the smog, Chicago had begun to electrify its public transportation, starting with the commuter lines that reached into the suburbs. Nonetheless, the *Chicago Tribune* ranked the city's top priority as "Lessen the smoke horror."

Hubbert had taken a roundabout route that summer, traveling more than a thousand miles from his home in Texas to reach Chicago. On arriving in mid-September, he planned to register for classes at the university, but on reaching the campus, he discovered its offices were still shut for the summer.

He then faced a more immediate problem. He had planned to ask the university for the address of the only person he knew in Chicago: Kenneth Luechauer, his former chemistry teacher as well as roommate. He had no way of finding his friend.

.   .   .

FOR THE PAST two years, Hubbert had studied at Weatherford Community College, a tiny school in Texas. Having grown up on a farm outside San Saba, in the Hill Country of central Texas, one of the most isolated and poorest parts of the nation, his family owned a small farm on marginal land but little more.

But his parents always placed a high value on education. His mother was a lot smarter than his father, and—unusual for the area—she had even attended junior college, forever after the high point of her life. She'd received that education thanks to the help of two men. One was a farmhand, a Mr. King, who repaid her family's past generosity by giving them the money for tuition for her to attend the school for a year. The other man was Francis Marion Behrns, head of the junior college Cora had attended. To honor these men, Cora named her first son Marion King Hubbert.

At age seventeen, King Hubbert had the opportunity to attend community college. He needed money for the train fare to Weatherford, 150 miles north of San Saba, so he sold his cow, one of his few possessions, for twenty-five dollars. Once at Weatherford, he'd worked his way through school, helping renovate the building and working odd jobs, from janitor to librarian. While there, he made an impression on his teachers as a quick learner. Inspired by his chemistry teacher Luechauer, his favorite subject became chemistry—and he decided he'd focus on that when he went on to a four-year college.

Hubbert also gained a reputation as a rebel. He was raised a Methodist, and his community read the Bible literally. When traveling preachers came through the area, the locals turned out their cattle to graze and camped out for several days, getting worked into a trance. He grew skeptical about this fervor, however. As he later put it, "I remember wishing that all this hocus pocus wasn't so." When a cousin told him about hypnotism, Hubbert decided that was what these traveling preachers were practicing.

The ones most susceptible to this kind of hypnotism, Hubbert thought, were the weak-minded, the dumbest people—and they often went on to become preachers themselves. His community, however, considered these preachers to have been called to that path by God—and Weatherford, a Methodist school, offered them free tuition. This irked Hubbert, who worked hard to pay his way through school, driven by his own fervor for science. When assigned to write an essay for an English class, he laid out his ideas on "the fallacy of being called to preach." Rarely did anyone declare such heretical thoughts so openly. Luckily for him, his English teacher harbored similar sentiments. Soon Hubbert was accepted into a small circle of skeptics at the school.

He made the library his "private study," staying late to work in peace. Toward the end of the school year, Weatherford's president, Fred Rand, came into the library one night and found him studying there.

"What do you want to do next year?" Rand asked.

"I don't know," Hubbert admitted. "I'm in a quandary." He had nearly finished his associate's degree and wanted to go on to a four-year university. But he also would need to work to pay his way through school.

"Why don't you go to the University of Chicago?" Rand said.

As Hubbert recalled, "I nearly fainted. I hadn't thought of going beyond the boundaries of Texas." Most universities in his home state seemed to have two dominant concerns, football and oil—neither of which interested him.

In contrast, Chicago, a thousand miles away, was one of the nation's premier centers of learning. From its start three decades earlier, it had been endowed by the oil magnate John D. Rockefeller to become "the Harvard of the West," albeit with a trailblazing attitude distinct from the Ivy League back East. Hubbert had heard about the school from local fundamentalist radio preachers. It was their prize hate, because its "liberal" religious scholars interpreted some episodes in the Bible, such as the Virgin Birth, as allegorical rather than literal truth. "That was my first intellectual tie to the University of Chicago," Hubbert recalled. "I had a sympathetic interest in Chicago because of this religious thing."

He decided he would make the journey to Chicago.

On leaving Weatherford in the summer of 1923, he was broke. With the help of a small loan from the school's president, he set out for Chicago, working his way there, first harvesting wheat in Oklahoma and

Kansas, then laying railroad tracks in Nebraska and Colorado. Along the way, he earned enough to get himself to his destination and figured he could find a job somewhere in the city once he arrived. "If you can't make it in Chicago," went a saying at the time, "you can't make it anywhere."

In mid-September 1923, Hubbert quit laying rails in Colorado and took the train to Chicago. His mentor and friend, Kenneth Luechauer, coming to attend graduate school in chemistry, had the money to travel straight there. So when Hubbert arrived and found the school closed, with no way to find his friend, he decided to start looking for a job.

He picked up a newspaper and scanned the want ads. He then headed into the Loop, the packed heart of the city. Locals boasted it held the densest cluster of buildings anywhere in the world, as well as the world's busiest intersection. The Loop's buildings were so tall they had been dubbed "skyscrapers," forming canyons packed with people and machines. There were crowded sidewalks and flashing signboards and automobiles whizzing around. Streetcars screeched and rumbled on the roads and on iron tracks elevated overhead. Hubbert, who'd never been in a city anywhere near as large, was overwhelmed and confused, his attention constantly pulled this way and that.

Chicago had more than 3 million residents—and yet as Hubbert walked along one of the crowded sidewalks, he ran into his friend. It turned out Luechauer had already gotten settled in at a boardinghouse in Hyde Park, just a couple of blocks from the university campus, and he invited Hubbert to move in.

THE UNIVERSITY OF Chicago campus was filled with foreboding architecture, the buildings constructed of gray limestone, their pointed window arches filled with stained glass. Clinging to the buildings' sides were elaborate foliage-like decorations and countless gargoyles, carved from stone and encrusted with soot.

A couple of weeks after Hubbert arrived, the university opened, and he went to register for classes. But as he recalled, "when they looked my credentials over, there was considerable misgivings." They would grant him only a few credits for his classes at Weatherford and would admit him only on probation. If his grades weren't good enough, he'd be dismissed. Worst of all, they wanted him to pay up front for a full year's tuition—money he didn't have. They suggested he forgo the fall quarter and work full time to save up what he needed.

Hubbert had been a tinkerer for years. His childhood hero had been the inventor Thomas Edison, the subject of one of the few scientific books in the one-room schoolhouse where he had started his education. Early in his teens, his sister Nell had bought him a prized book, *The Boy Mechanic: 700 Things for Boys to Do*, as well as a subscription to *Popular Mechanics*. Messing around with car ignitions and telephone magnetos, he'd tried to learn all he could about electricity.

So Hubbert sought jobs with telephone companies, landing a position as an installer for Bell Telephone. They sent him to the city's poorer areas, squalid neighborhoods in the southwest around the city's vast stockyards, where businesses dumped carcasses and other waste into the river, nicknamed "Bubbly Creek."

One day while chatting with a local who'd grown up on the streets, Hubbert got a tip: post office clerks got paid a lot better than telephone installers. In late 1923, just after his twentieth birthday, he secured a job as a post office clerk on the graveyard shift, earning $115 a month, about a third more than his salary from Ma Bell. But he needed to earn still more, so he also got a job as a waiter at a restaurant near his boardinghouse, where some of his housemates already worked. Between the post office and restaurant, he usually put in about thirteen hours a day, plus a two-hour round-trip commute by streetcar to and from the post office. Then the Christmas mail rush began, and the post office demanded twelve-hour shifts on top of his other work. Once he clocked nearly forty-eight hours straight, with time to catch only a few hours of sleep in between shifts and a meeting with the dean at Chicago. Although this work "nearly killed me," Hubbert recalled, "I hung on until I had the right amount of money."

In early 1924, with the start of the winter quarter, Hubbert was finally able to begin school, while continuing to work at the post office.

He knew his previous math courses had been poor, so he signed up for that subject. Luechauer told him German was the language of science, so he took that as well.

Growing up, Hubbert and his family had regularly attended church. But in Chicago he was bothered by what he saw as the church's ignorance, even in that relatively enlightened city. He quit going, but as a result he lost most of his social circle in Chicago. He was truly on his own and fell into a depression.

Nonetheless, he maintained interest in school. In signing up for his spring quarter courses, the biggest issue was not the workload but the wealth of options: forestry, geology, chemistry, and more. Inspired by Luechauer, he wanted to delve into chemistry, but it required long hours in the lab, which didn't fit with his work schedule. Instead, he took a geology course. He'd grown up hearing the biblical version of Earth's history, but also, he recalled, "I'd learned that geologists had a different chronology, and I wanted to know about it."

Through that geology course, he became friends with Frank Melton, the graduate student who taught it. The summer of 1924 Hubbert was working at a gas station and had extra time on his hands, so Melton recruited him to help with his doctoral project, reading tables of numbers and doing calculations. Melton was working on measurements of gravitational pull, in an attempt to discern geological features deep under the surface, grappling with a concept called isostasy, which occupied the center of a debate that had been running for several decades.

The prevailing idea was that Earth's crust—the planet's brittle surface layer—was some fifty miles thick. The crust floated on material beneath, which, under the effect of great heat and pressure, behaved like a viscous liquid such as honey. However, this picture contained a mystery: How could mountain ranges exist? All the additional rock that stood above sea level would require something below the surface to hold it up. A leading theory maintained that underneath mountain ranges, the crust was thicker. If true, mountains would be like icebergs, with part visible above ground but a large portion under the surface that helped keep the mountains afloat.

There was hope of testing such theories using new tools called torsion balances, which had only recently been brought to the United States. These sensitive but finicky instruments could pick up subtle changes in the gravitational pull from place to place—and potentially reveal the underground portion of a mountain range. But measurements from torsion balances were difficult to interpret, since no one could actually see the underground masses that were supposed to be responsible for the changes in gravitational pull from one place to another. Scientists had only a rough idea of how thick Earth's crust was and how its density might vary underneath plains, mountains, and oceans. So when Hubbert learned about the theories behind the gravity measurements, positing various unseen masses buried underground, he wondered, "was it real or was it not real?"

Melton's thesis didn't crack any big problem, but helping with it gave Hubbert his first taste of cutting-edge science. The basic skills involved—how to calculate gravitational forces, how to make sense of measurements of things you couldn't see, how to define problems clearly to avoid misunderstandings—would serve him well in later years.

AT CHICAGO, HUBBERT earned respectable grades but had no overall plan. Melton encouraged him to go for breadth, so he simply took whatever interested him the most. To him, this approach was ideal. But toward the end of his second year, a dean at the university scolded him: "You haven't declared your major."

"I don't want to major in anything," Hubbert replied. He had come to Chicago to get "an education," as he put it, not to focus on one subject.

"But you've got to," the dean insisted. "The university rules require it. Not only that, but you have to have a minor."

"Well, I may consider it," Hubbert said brazenly.

After this encounter, Hubbert studied the course catalog carefully in search of a way to satisfy both the school and his curiosity. He discovered an option for a joint major in geology and physics. For a minor, the obvious choice was math. Choosing this path, it turned out, would be easier than following it.

The school considered geology his primary major. Accordingly, his assigned adviser was the geologist J Harlen Bretz, a formal and unsmiling man in his early forties with a large bald dome of a head and a reputation for riding students. Instead of simply lecturing like all the other geology professors, in his classes Bretz put students on the spot by asking them tough questions, especially if they gave replies that were confused or nonsensical. One week he might pick on one student, the next week another, sometimes bringing them to tears.

Hubbert had yet to take a class from Bretz, but he'd heard the stories. Nonetheless, he wasn't intimidated. He went to their first meeting thinking, "I haven't committed any crime I know of."

Bretz marked down all the courses Hubbert would have to take—including Bretz's month-long summer field trip to Wisconsin, required for all geology majors. This news gave Hubbert a jolt. The field trip alone would cost him $120. But he'd planned to work all summer, aiming to save $80 a month for expenses for the coming school year. With this field trip, instead of starting the next school year with the money he needed, he'd have only "about $60 between me and starvation," he wrote to Nell.

Hubbert scrambled and found a job as cook on an archaeological dig in Wisconsin. It ran only a month but allowed him to be outdoors. He was surprised others weren't vying for this type of position. "Not one college fellow in a thousand even thinks about doing such work," Hubbert wrote to his sister Nell. But on the other hand, he added, "not one of those in a hundred has knowledge of outdoors enough to handle the work."

When his job as camp cook ended, though, he was almost out of money.

2

Pioneers' Eyes

BEFORE LEAVING FOR DR. BRETZ'S month-long field trip in Wisconsin, Hubbert had asked if there was anything he could read to prepare for it. "Not a thing," Bretz told him. "Not a thing."

Hubbert was short on money, but fortunately Bretz gave him tasks on the trip that spared him some expenses. The campsite was at Devil's Lake, two hundred miles northwest of Chicago, and Hubbert's first job was to drive up there in a small bus that the group would use during some of the longer day trips. "That was a godsend," he recalled, "for otherwise I'd have to buy train fare."

At the camp, Bretz ran things following long-standing traditions. At the start of each day, he stood outside the students' tents, banging on an iron barrel loop. If anyone failed to get up, he'd enter the tent, count to three, and dump the student out of his cot. Bretz also enforced a morning dip—stark naked—in Devil's Lake, which was frigid even in late summer, followed by a hearty breakfast of eggs, bacon, and toast. Then each day Bretz took a pair of students out for a field trip. The first day Bretz chose Hubbert.

Leaving their camp, which was in a valley at the southern end of the lake, they hiked up a gentle slope to the crest of a ridge that overlooked a canyon with five-hundred-foot cliffs on either side. Standing on the ridge, they had a view of the whole landscape around the lake. Bretz pointed out significant rock formations, such as some stone pillars that stood out, like the remnants of ancient monuments. "What are you looking at?" Bretz asked repeatedly. "What do you see?" He didn't instruct them, at least in any usual sense. Instead he asked question after question. Looking at the lake, he asked how it might have formed. The students guessed that a glacier had come through the area. But on further probing, they realized a glacier would have easily toppled the rock pillars they saw, so they reasoned that the ice must not have reached that point in the valley.

Hubbert reveled in this approach, in which they looked at the landscape through pioneers' eyes. "It was as if you were the first man in the area," he recalled. "You were just describing these things and wondering what went on here."

In camp each evening, Bretz quizzed the students to find out more about what they'd noticed, and how they might explain it. If they were confused, he sent them back to the same locale for another day of fieldwork. When not out in the field dodging rattlesnakes, they reviewed their notes, fleshed out their maps, and prepared for further hikes by replacing hobnails that the jagged quartzite rock had torn from their boots.

During nights around the campfire, they got to chat with Bretz—and there Hubbert discovered that his teacher's background was remarkably like his own. They had both grown up on farms, and Bretz had likewise begun to question religion at an early age, nursing contempt for the fervent believers around him.

These chats apparently had an effect on Hubbert. While still at Devil's Lake, he wrote a long letter to his sister Nell, the only one of his four siblings whom he felt close to. He admitted he'd quit church soon after arriving in Chicago—and as a result, he'd become lonely, "steeped in cynicism and pessimism." But he later moved into a new boardinghouse, where the other residents "restored my youth, taught me to dance, sing and be happy, and gave me an optimistic outlook on life again." Coming out the other side of this, he told Nell, "at present I am what would commonly be called an atheist." He was "tolerant of all religions—Christian, Mohammedan, Buddhist, alike, but follower of none. I like people and have a great deal of faith in them, in spite of all failings."

Exhilarated by Bretz's challenging teaching and with his religious skepticism strengthened, Hubbert was riding high. As he described to Nell, "I've been engaged in tracking the Great Glacier, finding Cambrian Trilobites, fighting modern mosquitoes, etc. Doesn't that sound interesting?"

Within a few days, Hubbert's newfound optimism would be tested. Bretz had waterproofed the camp's tents using a common method, taught by Boy Scouts: painting the canvas with a solution of paraffin dissolved in gasoline. When the gasoline evaporated, it left behind a waxy layer—but also made the tents dangerously flammable. One night Hubbert and a handful of other students were working in a tent, using a small stove to keep warm. Suddenly their tent caught fire. They escaped but lost all their possessions, except the notes in their hands. Hubbert had owned only two suits to start with, and one burned up, leaving him with the clothes on his back. To get through the frosty night, the tentless students broke into a closed hotel nearby and sandwiched themselves between mattresses to keep warm.

After the mishap, Bretz called an early end to the trip—but not before holding a final assembly. "We've found out quite a bit for this course, and it's completed," Bretz told his pupils. "You've done the work, you've seen the evidence, and you know what went on here." Then he suggested, "Why don't you write a report on it for another half credit?" Hubbert didn't look forward to writing the paper, but he needed the credit, so he signed on.

"When you write your report," Bretz continued, "I want a complete bibliography of the earlier papers on this area. You are to read those first, before you start to write your own report." This was the exact opposite of Bretz's approach before the field trip. Earlier, he'd wanted them to read nothing—but afterward he wanted them to read everything.

3

Bugged Out

BACK IN CHICAGO, THE FLURRY of fieldwork over, Hubbert had to sort out his finances. Before the fire, he'd figured he would have to ask the school for a loan of about $200, enough to cover tuition and books. But after the fire, he was in a much worse position. Someone advised him to ask the president's office for help, and, lucky for him, the administration dug up $250 for a grant, based on need rather than achievement.

Hubbert soon got another piece of good news: he'd landed a job as a lab assistant in geology. During the field trip to Devil's Lake he'd apparently impressed Bretz, who'd recommended him for the job, putting him in charge of a few other student assistants. This meant the end of his grunt work—no more laying rails and harvesting wheat, no more waiting tables and cooking in camps.

Around this time, Hubbert went to the registrar's office to review his progress. Although he knew Chicago's courses were of far higher quality than those he'd taken at Weatherford College, he hoped to secure more credits for that earlier work. The registrar, a woman of about sixty, pored over his records intently. "Now, here's this credit in agriculture," she said. "You know, these small schools, they just don't have adequate laboratories."

Hubbert was flabbergasted. "I was raised on a farm," he replied. "I displaced a farmhand from the age of ten, working with plows and horses. Do you mean to tell me Chicago students have that much laboratory in agriculture?"

She admitted they didn't. He got the credit, putting him one step closer to his degree.

That fall Hubbert immersed himself more deeply in science. A year of advanced geology covered the whole scope of the field, taught by Bretz together with a well-known geologist, Rollin Chamberlin, whose father had founded Chicago's geology department. Crucially, the students got the task of reading many classic papers in geology. "Before with literature as we'd been brought up in high school, authors were just an abstract 'they,'" Hubbert recalled. "Here all of a sudden they became living people—human beings, fallible human beings."

Over Christmas break, Hubbert finally found time to tackle his extra-credit report about the trip to Devil's Lake. Instead of being drudgery, as he'd expected, writing the report was a revelation. He discovered how scientists gradually built up their understanding, layer after layer, like geological strata. But sometimes there was a rupture, where new ideas erupted forth. Most important, he learned that some old, cherished ideas—even from eminent geologists—could be simply wrong.

One example was a rock formation Hubbert had seen during Bretz's summer field course, a chimneylike pedestal about twenty feet tall with a seam running vertically down the middle. One half was a lighter quartz rock, the other a dark, clay-rich rock. These had once been two layers, one sitting atop the other, and over the tumult of time had been tilted on end. Then the neighboring rock had eroded away. The unusual rock bore a bronze plaque placed by students a couple of years earlier, commemorating this as the "Van Hise rock." To the students on Hubbert's field trip, the rock stood out, holding some important clue about the area. But how it was formed, and what story it told, Bretz had forced the students to puzzle out on their own.

Back in Chicago over Christmas break, Hubbert read through the reports on the region. It turned out that Van Hise had, in his initial report, gotten the layers of rock in the area upside down, mixing up which were oldest and which youngest—something later geologists figured out.

It was ironic, Hubbert thought: "The students of the University of Wisconsin put a plaque on one of Van Hise's major boners."

From childhood Hubbert had always been taught—at home, in school, in church—that books were authoritative. If you read something in a book, you could trust it. At Weatherford, he'd begun to question this attitude when it came to religion. Bretz's course also made him question, in a much more systematic way. Researching and writing his report on the field trip, he learned to reserve judgment, to become convinced only when he'd seen sufficient evidence.

It was "one of the major intellectual experiences that I've ever had," Hubbert recalled. In a tribute to Bretz decades later, he wrote, "We learned that the surest way of getting ourselves hopelessly enmeshed in other people's errors is to read the literature first." Instead, the first step should be to "get our hands dirty with the data of the problem itself. Only after that is it safe to read the literature, and even then with extreme caution and principally for data. Data are usually reasonably reliable; the author's interpretations are where the danger lies."

IN HIS ADVANCED geology course, Hubbert was puzzled by the terminology. "My God it was a hodgepodge if I ever saw one," he recalled. "Every kind of a screwball name for things they could think up." There was a zoo of names for different kinds of faults: strike faults, bedding faults, dip faults, oblique faults, peripheral faults, radial faults. Each type of fault spawned variations, generating compound names like "reverse dip-slip faults."

Hubbert wasn't the only one who found it perplexing. A decade before, the Geological Society of America had set up a special committee to sort through the terminology and clean it up. What the committee called "strike-slip faults," for example, had been dubbed by one earlier author "heaves" and by another "horizontal faults," while a third author used two distinct names, "transcurrent faults" and "transverse faults." Despite the muddle, the committee's recommendations were cautious, in one case stating that a particular change of names was "impossible" because "miners would never accept it." The recommendations, published as a pamphlet and meant to serve as a practical handbook, remained a maze of terms. For all the committee's efforts, it had done little to remedy the situation.

In the geology course, Hubbert was assigned to review this committee's report. In his presentation to the class, he followed Bretz's approach to exposition. "I reviewed it by asking them questions. 'What does this mean? What does *this* mean?'" Hubbert recalled. "Just to demonstrate that the thing was a mess." Instead of tinkering with the existing terms, he argued, they should start from scratch, analyzing faults in terms of their fundamentals. "Why don't we just reduce them down," he said, to their "elemental, geometrical components?" A fault was simply the meeting place of two slabs of rock, in the midst of gradually sliding past each other. It would be far better, Hubbert argued, to describe each type of fault in terms of three basic properties: the horizontal displacement, the vertical displacement, and the angle. Along with these measures, a sparse list of shorthand names would suffice to cover every type of fault.

Hubbert's teacher, Rollin Chamberlin, was immensely impressed with this report and said he'd like to publish it in *The Journal of Geology*, which he edited—one of the top journals in its field. Once the paper was formally accepted, Hubbert, still an undergraduate, became a published scientist. But the journal didn't reveal his status. When the paper appeared, it listed his name and affiliation simply as "M. King Hubbert, University of Chicago."

IN THE SPRING quarter of 1926, Hubbert took the final courses to complete his degree. Two he signed up for only grudgingly, but they both sparked new interests. One was a surveying course taught by a geologist from a relatively new oil company called Amerada, which used these courses as a way to scout for promising students. Hubbert's work soon earned him a job offer, and he thought it a good position, with good pay. He intended to continue on toward his doctorate but took the position for the summer.

Before he earned his degree, though, he also had to complete a course on economic geology—the study of minerals useful to humanity, for industry or agriculture. It was taught by Edwin Bastin, an oaf of a professor who stood a head taller than the rest of the faculty and wore his necktie short, reaching only the top of his paunch. "I didn't think too much of Bastin," Hubbert recalled. "I thought he was a little bit on the dull side and not too bright." But after attending Bastin's daily lectures, he realized the professor really knew his material, and the course was "beautifully organized." Bastin, it turned out, had been in charge of the mining and metals branch in the US Geological Survey before the Great War, and during the war he had helped the government locate the resources it needed to build the ships and trucks, guns and tanks for a modern military.

For homework, Bastin had the students comb through thick books of statistics, gathering data and plotting it out on graphs. In some assignments, they traced the history of production of the local steel and cement plants over the years. In others, they studied the main categories of railroad freight, which were mainly the products of mining. "This was just a complete eye-opener to me," Hubbert recalled.

Chicago was perhaps the most industrialized city in the world, a sharp contrast to Hubbert's rural childhood. Yet, he recalled, "it had never dawned on me that this modern world runs on minerals." In geologists' parlance, "minerals" included not only metals and rocks but also the fossil fuels that powered machines. The industrial world was, in essence, built of iron and fueled by coal. But machines didn't merely consume fuel. They also helped pull it from the ground and transport it to where it was needed. Machines and fuels were symbiotic, the development of one furthering the other.

When Bastin showed the class a huge wall chart on the history of coal, Hubbert experienced another revelation. It showed how coal production had grown exponentially, with more and more extracted each year. "My eyes bugged out," he remembered. Bastin explained that, in theory, coal extraction could continue growing as quickly as it had been, until the known coal reserves were suddenly all gone. In that case, a simple calculation showed that the coal would be used up in about a century. But that scenario—with rapid growth up until the moment the final lump of coal was burned—was unrealistic. They also considered another situation. Instead of consumption soaring, if it held steady at the current rate until all that coal was exhausted, the known coal would last six thousand years.

Hubbert sketched out these cases in his notebook, along with a dotted line for a third possibility, between the two extremes—the possibility that coal consumption would continue rising but at a slower rate. The reserves would last longer than 100 years but less than 6,000 years. There were limits in sight. Also, not all resources were equal. As Hubbert wrote in his notebook, "Within 50 years best coals may be exhausted."

Intrigued by this issue, Hubbert went back to the original data and drew the points on a particular kind of graph called a semilogarithmic plot. When a quantity grows by a constant percentage each year, its history forms a straight line on a semilogarithmic graph. Hubbert plotted the points for coal, year after year, and found a fairly straight line that persisted for several decades: a continual growth rate of around 6 percent a year. At that rate, the production doubled about every dozen years. When he looked at this graph, it was obvious to him that such rapid growth could persist for decades—his graph showed that had already happened—but couldn't continue forever.

A few lectures later Bastin covered petroleum. "Very young industry," Hubbert noted. "Started 1859." He copied down a list of statistics for annual US oil production, showing its spectacular rise. During Hubbert's lifetime—the past two decades—production had risen more than sevenfold, from 100,000 barrels a day in 1903 to 760,000 barrels a day in 1924. "How long will it last?" he wrote. "Difficult to estimate reserves."

"That course was one of the most revolutionary courses I ever had in my life," Hubbert recalled more than sixty years later, at age eighty-five. These questions of when growth might hit limits, and how to gauge remaining resources, would occupy his thoughts for decades to come. "I was very strained over this thing," he added, "and I still am."

Complete Rebellion

"NEW OIL TOWN IS A Modern Sodom" blared a typical headline about Borger. Little more than a single, unpaved street on the Texas Panhandle near Oklahoma, Borger was packed with hastily erected, flimsy structures of wood and corrugated sheet iron. It featured boardinghouses and gambling houses, dance halls and brothels, and—flouting Prohibition—a variety of saloons. At night, fights often broke out in the streets, making outsiders wary of venturing out after dark. Summer rains turned the main street to mud deep enough to regularly engulf Model T's and even trucks. Lacking running water or sewage, everyone used pit toilets. The town became so overrun with rats that the movie house, the Rig Theatre, issued a bounty: for a free admission, bring in ten rat tails.

Just months before, Borger had sprung up from nothing, to house and serve the oil field workers. It was where Hubbert wound up in his summer job for Amerada Oil, working on "the biggest, wildest developing oil field in the country," as he put it.

A DECADE BEFORE the oil boom, a geologist had scoped out the area and told locals, "No one knows whether or not the Lord has put any gas or oil in the Panhandle of Texas." He chose a spot to sink the first well in the area, on a broad uplift of the land dubbed the John Ray dome. "If there should be any oil or gas in this part of the world, this would appear to be the best place to find it," the geologist said. He was right. They didn't find oil but did hit gas. The wells proved so prolific that the Panhandle field came to be called "by far the world's greatest gas field," reputedly holding "an inexhaustible supply of fuel."

Then in 1925, near the town of Amarillo, came the area's first big oil gusher. A well, on encountering a reservoir nearly 2,200 feet down, became a geyser of petroleum, launched into the air by the high pressures underground. This discovery set off a drilling frenzy in the region. New wells came rapid fire, week after week, as crews drilled wherever they could. Suddenly drilling derricks were everywhere—towers of timber or metal, erected to hold lengths of pipe as they went down the hole. Like tenacious plants, the derricks sat "on top of a towering knoll, on the ledge that juts from the side of a canyon's wall, on the shoulder of a rock-strewn hill," a visiting journalist noted. Production rose quickly, and within a few months it was clear that Amarillo was one of the nation's biggest oil finds to date.

By the spring of 1926, the oil was coming so fast that companies were having trouble handling it. There weren't enough steel storage tanks, so they built earthen berms to hold huge pools of oil. There were no pipelines to the area yet, so all the oil had to travel by rail—and its transport got held up when the railroads gave priority to hauling the summer wheat harvest. Texas had a body, the Railroad Commission, that regulated railroad freight and that had begun reaching into regulating oil production as well. To temper this boom, the Railroad Commission banned "shooting" of wells, the practice of dropping nitroglycerin bombs down the hole to break up the rock and boost production rates. But producers complained about the ban, and the Railroad Commission backed down. The boom continued.

In the oil patch, there was no regulation over how much you drilled or how much you let gush out of a well. Competing companies were all tapping into the same layer of rock, which was cut through with fractures and channels that allowed the oil to flow through the rock. If one company held back, another on a neighboring plot would simply drain the oil itself. So it was in each company's interest to drill as fast as possible, even if the collective result was a glut that held prices down and undermined profits.

Hubbert spent the summer working on crews that surveyed hundreds of oil wells across three counties. He realized the important role oil played in society, but still, when he watched the way the drillers went about their business, he wondered about the sanity of it all. Many found the gold-rush atmosphere exhilarating, but Hubbert thought it was "like so many buzzards fighting over a carcass."

ONCE THE SUMMER was over, Hubbert headed back to Chicago to begin graduate school. On top of his coursework he juggled three odd jobs—including a dollar-a-day task as a seismograph tender, overseeing an earthquake-detecting device that had to be handled like a baby. All this kept him busy from six a.m. to eight-thirty p.m., with almost every hour blocked out. But a combination of curiosity and stubbornness led him to take on even more.

In Rollin Chamberlin's graduate-level course on structural geology, Hubbert hoped to learn more about how mountains built up. In an undergraduate geology class the year before, Chamberlin had given an explanation for the folds in the Appalachian Mountains. But it hadn't made sense to Hubbert, so he'd asked for more information. At the time Chamberlin had deflected the question, telling Hubbert to wait for the graduate-level course. When the formation of the Appalachian Mountains came up in the graduate course, Chamberlin again gave an unsatisfying explanation. Chamberlin said that in that area, the folds were almost parallel to the forces compressing the mountains, whereas typically such folds formed at right angles to the compressive forces. But in that case, Hubbert figured, forming the Appalachian folds would have required forces of almost infinite strength—an unrealistic situation. Something must be wrong with Chamberlin's explanation, Hubbert thought. The two of them "got into a great hassle over that," as Hubbert remembered. There was "almost ill feeling between us. I finally decided that I couldn't talk with him."

Hubbert sought a solution on his own. He had access to a room in the basement of the geology building, Rosenwald Hall, where he built an apparatus to test his ideas. First he sawed a big block of wood in half and packed the gap in between with modeling clay to simulate rock, which could deform when put under high pressures. Then he pressed a lattice-work of strings into the clay and painted circles on the surface of the clay, which gave him a way to monitor how the clay responded as he slid the blocks of wood and compressed the clay.

While Hubbert was in the midst of building this contraption, a fellow student mentioned it to Chamberlin, who then called Hubbert in.

"Why didn't you tell me about it?" asked Chamberlin. He had done a lot of work on such scale models.

"Well, I'm not finished," Hubbert explained. "I'm just trying to find out something. And I don't have anything to discuss."

When Hubbert had his results ready, Chamberlin offered him an hour at the weekly meeting of the school's geology club to present his findings—giving the other students an opportunity to "see if they could tear it to pieces," as Hubbert put it. Chamberlin attended as well.

Hubbert described his experiment, which suggested that the forces at work must have been pointing in a different direction than Chamberlin had thought. No one found any big holes in Hubbert's approach. At the end of the session, Chamberlin was almost apologetic. Hubbert hadn't needed to do all this alone, he said. He could have helped Hubbert out.

"Well, it was a simple experiment," Hubbert demurred. "It's done, and it doesn't mean anything else."

But Chamberlin took it more seriously. He offered to print the results in *The Journal of Geology*. With this, Hubbert landed his second scientific publication. "I've developed an awful case of the big-head," he confessed to Nell. "Imagine all the future classes of this great department having to read my stuff along with the other geological classics."

However, doing these experiments and writing a paper, on top of his jobs and coursework, Hubbert was burned out again. "I needed a little change," he recalled, "a breath of fresh air and some cash." He signed up with Amerada to spend fifteen months working in the oil patch.

IN THE SUMMER OF 1927, Amerada put Hubbert on a "geological party," a crew that traveled around Texas, New Mexico, and Oklahoma, surveying and mapping areas that might hold oil. Although the work was fairly mundane, it gave him an opportunity to learn more about oil exploration. After about six months, he was invited to join a sister organization called Geophysical Research Corporation (GRC), launched a couple of years earlier.

In the oil business, GRC was a pioneer in the use of geophysical devices. The company tested torsion balances that picked up subtle variations in Earth's gravitational pull, and magnetometers that did the same for magnetic fields. They also tried seismographs. Rather than sitting in a university and detecting natural earthquakes, like the seismographs Hubbert had tended in Chicago, GRC's devices were brought into the field to listen for subtle man-made quakes from blasts of dynamite. Each explosion sent sound waves traveling through the ground, which the seismograph could pick up at a distance, revealing underground rock formations that would otherwise remain invisible.

Hubbert worked on crews that were experimenting with two competing approaches. One, called a refraction seismograph, could detect small variations in the speed of sound waves, depending on whether they traveled through the typical rock in the area or through some unseen anomaly. In particular, geologists hoped to find salt domes, huge plugs of pure salt that rose up from the depths. Since the salt was less dense than rock, it would push rock aside over millions of years, forming a tall column of salt. And since oil couldn't flow through the salt, it could get trapped along the sides of the salt column. The oil industry looked for signs of salt formations where they reached Earth's surface, pushing up the ground and creating a dome. The world's most prolific oil field to date—Spindletop, outside Houston, Texas—had been found on the flanks of a salt dome.

Spindletop's discovery in 1901 set off a drilling frenzy on every visible salt dome, and by the 1920s the oil business was looking for ways to reveal additional salt domes that might be hidden beneath the surface.

Refraction seismographs had uncovered some salt domes along the Louisiana coast, but the technique worked only for showing relatively large-scale features. A competing method, which GRC helped develop, was called the reflection seismograph. It required more subtle measurements, but promised to reveal finer detail. Instead of looking for differences in the speed of sound waves that traveled through an anomaly, it looked for faint echoes that bounced off the various layers of rock underground. The effect was like looking through a window, where you can often see a ghostly image of yourself reflected back. Although not a mirror, the glass still reflects back a small portion of the light that hits it. Similarly, where two different types of rock meet underground—such as shale and sandstone—sound waves traveling across that boundary would get partially reflected.

Hubbert's crew traveled across Texas and Oklahoma, testing out both types of seismographs. In early January 1928, his party went to Norman, Oklahoma, where Frank Melton, his good friend and former teacher at Chicago, was a professor at the University of Oklahoma. Hubbert was one of the most promising young scientists in the whole country, Melton told him, and Melton tried to line up a job for him as an instructor at Oklahoma.

While waiting to hear about that possibility, they began work on a research paper. This time, instead of student and teacher, the two men were collaborators. Hubbert's regular job kept him very busy, but he squeezed in time with Melton. "We only had one surveying crew and instrument," Hubbert recalled. "I trained one of my assistants to run the instruments, and so I would work maybe three or four days a week, from daylight to dark." That gave him more days off to brainstorm with Melton about how to improve the use of gravity measurements for finding oil.

GRC had also been using the torsion balance to take gravitational measurements in search of local anomalies that might form oil traps. Although torsion balances had managed to uncover some salt domes, in general the results had been mixed and not terribly promising. But Hubbert thought that as yet, the petroleum companies didn't know much about the complicated endeavor of interpreting the instruments' readings. In order to see the true effects of local features such as salt domes, they had to carefully correct for the gravitational pull from the surrounding landscape, such as mountain ranges or ocean basins.

Melton's thesis on isostasy had dealt with exactly this issue, looking at interpretations of gravitational measurements to infer what geological structures lay underground. He and Hubbert reviewed a few different methods of interpreting the results, some of which required extensive calculations and a couple of other methods that were greatly simplified. One of the simple methods was of no use, they argued. But there was an approach that worked well enough that it could reveal the relatively small geologic structures that might trap oil, using calculations easy enough to be practical to do in the field. Hubbert and Melton wrote up a paper and submitted it to the *Bulletin of the American Association of Petroleum Geologists*, which promptly accepted it. Hubbert had his third published research study.

GRC was impressed with Hubbert and offered him a permanent job. "It would have meant that I was in on the ground floor of this company," Hubbert recalled. Although the work was routine, he enjoyed it. "For that year, I didn't think I was wasting my time. I thought it was a damned good experience." But for the longer run, he said, "it wasn't what I wanted to do."

Despite Melton's efforts, no job offer materialized from Oklahoma. Hubbert did get a job offer from Texas Tech, in Lubbock, but he decided to return to Chicago to get his doctorate. In the meantime, in May 1928, he'd have to take the train back to Chicago to sit for the master's degree exam. "I haven't looked at my stuff for a year and shall start cramming about a day before the exams so I may flunk," he wrote to Nell shortly before leaving for Chicago. "At any rate I'll have about ten days of freedom in which to visit my friends and take in the shows."

He did pass the test. With his newly published paper on stresses in faults, based on the model he'd built in the basement of Rosenwald Hall,

Chicago awarded him a master's degree. He then rejoined his crew in east Texas and wrapped up his final few months of work in the oil business.

ON RETURNING TO Chicago in the fall of 1928, Hubbert resumed his old routine, once again taking courses and picking up a bit of money for tending seismographs. With a master's degree, though, he was able to get work teaching at Chicago. Meanwhile he searched for a focus for his doctoral thesis. He was studying seismology, gravity measurements, isostasy, and more—"quite a series of geophysical things," he recalled. His fieldwork had given him valuable practical experience with geophysical measurements, but he wanted to delve into the theories that would allow him to interpret these measurements. To better understand Earth's depths, he wanted to use rigorous analyses of forces, laid out in mathematical equations. "I began to stew more and more over the fact that nothing of that sort was being given in the geology department—or in the university," he remembered.

Around this time, he encountered the ideas of William Thomson, a famous British physicist who had been the first scientist named to the House of Lords—and thus was known as Lord Kelvin. To many Brits at the time, Kelvin was the public face of science, and he held a stark view of what counted as true understanding. In an 1883 speech, Kelvin declared:

[W]hen you can measure what you are speaking about and express it in numbers you know something about it. But when you cannot measure it, when you cannot express it in numbers, your knowledge is only of a meagre and unsatisfactory kind. It may be the beginning of knowledge, but you have scarcely, in your thoughts, advanced to the state of science.

To Hubbert, this rigor was seductive. As he later joked, during this time Kelvin was his patron saint. He grew disdainful of merely descriptive geology, wanting to stick only to what could be quantified.

In following this path, Hubbert felt his advisers at Chicago were of little help. He'd already had clashes with Chamberlin and didn't think much of Chamberlin's grasp of the physics underlying geological processes. Bretz had penetrating intuition for teasing out the history of geologic formations from examining the layers of rock and erosion patterns—including a controversial theory he developed that a catastrophic flood, long ago, had carved out deep canyons in Washington State. Although the evidence was on his side, Bretz's theory faced staunch opposition, including from the nation's experts at the US Geological Survey. Bretz was an inspiration for Hubbert, but was of little help with applying physics and mathematics to geology.

Hubbert thought the geology department as a whole had a narrow conception of geophysics, limited to devices that measured Earth. "Geophysics is not these instruments, it is not gadgets for finding oil," as he later put it. "It's the physics of the earth. Whatever's running downhill is geophysics, mountain-making is geophysics." Geologists studied these processes, of course, but with only a nod to the underlying physics. "Usually their notions of stresses and things of that sort were mostly erroneous," Hubbert said. "I was entirely on my own, just trying to dig these things out bare-handed."

Hubbert found some helpful books on the theory of geophysics in German—which, fortunately, he could read, having studied the language on Luechauer's advice. He and a couple of other students started to work together, trying to make sense of the research to date, which went far beyond what Chicago's courses covered. They were mounting a "complete rebellion over the state of geology," Hubbert thought, and he served as the "ringleader."

After several months of study, by early 1929 Hubbert decided he knew enough to teach a class on geophysics. Although it was unusual for graduate students to instigate new courses, he received permission from the geology department. "I am opening up a new field in which there are no teachers here," Hubbert wrote to Nell. "This concession on the part of my department toward me is a quite an unprecedented thing, and if I put it over it can't fail to have very significant and probably far-reaching consequences."

When Rosenwald Hall emptied as the geologists embarked on summer fieldwork, Hubbert returned to his solitary pursuits. He spent most of his time designing his upcoming geophysics course, while bringing in money with odd jobs, including flying weather balloons off the roof of the physics building.

IN THE FALL, on October 24, 1929, the stock market plummeted—a day immediately dubbed "Black Thursday." Stocks had been rising strongly for several years. The crash was a surprise, the topic of many newspaper headlines—yet most saw it as only a temporary setback. In early November, *Business Week* declared that the stock decline, while sharp, "doesn't mean that there will be any general or serious business depression." Over the coming weeks, stock prices continued to drop, but many thought a rebound was near. In mid-November Winston Churchill, until recently Britain's chancellor of the exchequer, received a cable from a friend, the wealthy American financier Bernard Baruch, that stated simply, "Financial storm definitely passed."

BY EARLY 1930, it was finally time for Hubbert to teach his geophysics course. It went well from the start, attracting one of the highest enrollments of any graduate class in the department. In late February, about two-thirds of the way through the quarter, Hubbert got a call to come to the geology department's office to meet a professor visiting from Columbia University in New York. It turned out Columbia also wanted to start offering geophysics courses, so Hubbert chatted with the professor for about an hour, then extended an invitation to his lecture that day. Over the following weeks, Hubbert exchanged letters with the head of Columbia's geology department, who invited him to come for an interview and to give a lecture there. During his spring break, he made the journey east.

Soon after Hubbert arrived back in Chicago, he received a job offer to lead a new effort on geophysics at Columbia, which had the largest geology program of any university in the country. Hubbert, then only twenty-six, told Nell:

> All this has happened fast enough to make one's head swim. I am to be teaching graduate courses in geophysics and structural geology—my two favorite subjects. I shall not be going on as an understudy of some old and famous man, but as the high muckety-muck himself with a clear track ahead. They expect me to make my division one of the best in the country within ten years (and I believe I can do it).

Columbia began paying Hubbert that fall, even though he would remain in Chicago for another semester, in hopes he'd finish his doctoral research. However, he was floundering on his thesis. He'd tinkered with geophysical devices and tried to build an electronic seismograph. "I was a general neophyte on what I was trying to do," he recalled, "and I didn't get anywhere." It didn't help that he was tired of school. "I have had about as much uninterrupted schooling as I can stand," he told Nell. "What I think would do me the most good just now would be to get a job as a ditch digger."

Nonetheless he was happy—and feeling flush. Between his regular earnings and advance pay from Columbia, he brought in $320 a month. "I used to think people had to work for a living!" he told Nell. He had his own office and a borrowed Victrola, and he splurged, buying himself a new record, Haydn's *Surprise Symphony*. In the nearby kitchenette he could brew tea, making his office a place to invite friends over to listen to music and have "bull sessions." He could even manage visits from young ladies, as long as he kept them brief. As he bragged to Nell, "It is really terrible to be a gay young bachelor."

Several months later, in the summer of 1930, he wrote to his sister about his love life:

> Have busted with Cora Louise and have lost most of my other girl friends from sheer neglect. Have a Russian friend, Yelena Pavli-

nova, in bacteriology who is a dear—brilliant as a whip! But she has gone into hibernation to study for her M.S. exam so there is little left for me to do but work.

Over the coming months, even as he planned to leave Chicago, he grew closer with Yelena—and she apparently grew fond of him, too. As he told Nell, when Yelena heard he and another girl had been joking about eloping together, Yelena "dragged me into the bathroom, shut the door, and accosted me with 'I hear you are getting married.'" She gave him a lecture about taking marriage seriously.

He enjoyed his final months in Chicago but did sound a somber note in writing to Nell. Their brother Leo had come up from Texas and was living with King but was having a hard time finding a decent job. For months, the economy had been sinking deeper into a depression. Given this, King thought his other brother, Fred, should stay put rather than come to Chicago in search of work. "It seems that the bottom is just about out of business here," Hubbert said. "Just now is a mighty dull time to look for jobs."

Nonetheless, he was hopeful about the future, saying, "Things of course will perk up sooner or later."

PART 2

New York City and Washington, DC, 1930–1943

The Meeting

"I HAVE BLUNDERED INTO A goldmine," King Hubbert wrote to Nell in early 1931, soon after arriving in New York City. "I now see it as the most important occurrence to me since I have been here."

A few weeks earlier Hubbert had climbed up the stairs to the Meeting Place, a club above a speakeasy called Lee Chumley's, in the heart of Greenwich Village. Unlike the rest of Manhattan, with its streets locked into their famous grid, the Village had retained its old twisted lanes, which slowed traffic and muffled the din of the wider city. There the Meeting Place occupied a spacious loft with vaulted ceiling, split up into small sitting rooms, each with its own fireplace, where visitors were often busy chatting and debating.

Hubbert had been invited there by a secretary at Columbia, and he soon discovered the Meeting Place was a haunt of the city's top intellectuals. The president, Max Eastman, was a radical publisher who had challenged the government during the Great War and wound up serving jail time for sedition. One regular, the Arctic explorer Vilhjalmur Stefansson, had a table permanently reserved downstairs in Chumley's, where he'd participated in an experiment in which he ate nothing but meat for a year. Many of the Meeting Place's visitors were writers, including Stuart Chase, who regularly covered economics for major magazines like *Harper's* and *The Nation*, as well as novelist Sinclair Lewis, the latest winner of the Nobel Prize in Literature. Soon after the visit, Hubbert wrote to Nell, "I was so fascinated with some of the members (and they with me!) that I was promptly elected to membership."

Hubbert was most fascinated by the man he'd been invited to meet, Howard Scott. Six foot five and broad-shouldered, with an enormous head, Scott had a rich baritone fit for the radio. He dominated any room. Fond of eclectic monologues, he could rattle off statistics on factory production as easily as he cited ancient history. Between incessant puffs on cigarettes, he'd drop wry quips like "A criminal is a person with predatory instincts without sufficient capital to form a corporation."

Over dinner, Scott told Hubbert of his wide-ranging work. During the Great War, Scott said, he'd been down in Alabama, serving as the technical director of Muscle Shoals, a huge plant that churned out nitrates—key ingredients for fertilizers as well as for bombs. In the decade since, Scott had been engaged in a largely solitary pursuit, developing what he called the Theory of Energy Determinants.

OVER THE MILLENNIA, Scott pointed out, mankind had continually devised ways to harness more and more energy. The control of fire, the advent of agriculture, the domestication of animals, the construction of windmills and waterwheels—all served to bend more of the world's flow of energy toward human ends. Another major step came with the use of coal to replace wood. At first coal was used primarily for heating buildings and for smelting iron. Then during the 1760s, Scottish inventor James Watt devised crucial improvements to steam engines that truly kicked off the age of coal.

A century later came drilling for oil. It was a superior alternative. It burned cleaner, making it ideal for lighting, and it packed energy more densely, enabling the automobile and the airplane. This marriage of machinery and fossil fuels had completely transformed how humanity used energy—yet society hadn't caught up with this revolution. "We have before us," as Scott put it, "the spectacle of a company of persons trying to run a social system under rules which actually were canceled on the day when Parliament confirmed James Watt in his patent on the steam engine."

Similarly, Scott argued, standard economics was obsolete. It held that the scarcer something is, compared with people's needs and wants, the higher the price should be. All of America's systems—for running government, creating money, and measuring worth—were based on such scarcity. Scott called this "the price system," an arrangement he saw as outdated. All manner of machines—cotton harvesters and flour mills, cigarette-rolling machines and steel plants—had developed to the point that they could produce a bounty, requiring people to work less and less to produce everyone's daily needs. This trend would continue, throwing most people in existing industries out of work. The scarcity that was the basis of the price system was being erased by the power of technology.

This process was nothing new, but Scott argued it was approaching a breaking point. So far, the price system had survived by encouraging various kinds of waste. Factories manufactured razor blades that quickly went dull. General Motors pioneered "dynamic obsolescence," constantly redesigning their cars. And there was "conspicuous consumption," in which owning—even wasting—became a status symbol. Through the 1920s, industry, investors, and consumers alike had bought more than they could afford, piling up debt. The Great Crash of the stock market proved how precarious the system had been, Scott argued, and he expected it couldn't last much longer. Guided by his Theory of Energy Determinants, for years Scott had been warning of a coming collapse. He admitted, with false modesty, that he'd been a bit off. He said he'd predicted it would come in April 1930—then the Great Crash struck in October 1929, several months early.

The way ahead, Scott argued, was to embrace the abundance created by machines. Even before the crash, industrialized nations had been gradually adjusting to the spread of mechanization by cutting down their standard work hours, from over eighty hours a week to below sixty. After the Depression struck, a share-the-work movement had sprung up, proposing to cut work hours further, to forty hours a week or even thirty, while requiring that employers maintain weekly wages about the same level. W. K. Kellogg had done it at his cereal factory with great success.

But Scott thought such plans didn't go far enough. He proposed that if the nation managed its factories efficiently—running them around the clock, churning out a standardized set of essentials—then workers would only have to put in sixteen hours a week. Everyone would have plenty of goods as well as lots of free time. The only thing in the way of achieving such abundance and leisure, Scott argued, was the price system—so it should be scrapped.

Scott envisioned government-owned industries, managed by scientists, engineers, and technicians. They would abandon money—no more dollars and cents—and instead would use "energy certificates." Everything from hamburgers to trousers would be paid for in terms of the energy required to make it—and that meant industrial energy, primarily from fossil fuels and hydroelectric power. Each citizen would earn the same allotment of energy certificates, as long as they put in the required sixteen hours of work each week, and it would be more than enough to meet each person's needs. The certificates couldn't be traded, and they'd expire after a year, so no one could accumulate wealth. Recording everyone's purchases, government-run industries would know exactly what they should produce in the future, and how much energy would be required to keep factories humming. Everyone would be equal, all needs met, and waste eliminated.

Scott had been developing this vision for more than a decade. When the New York *World* interviewed him in 1921, he'd described his overarching goal: "To figure out what the American people want, and get it for them."

AT THE TIME Hubbert met Scott, the Depression was growing worse and unemployment continued to climb. Yet optimistic slogans were in abundance. Prosperity was "right around the corner," and soon everyone would have "two chickens in every pot, and two cars in every garage." Such promises never materialized, however, and none of the nation's leaders had a credible explanation for what had gone wrong in the first place. Even more vexing was crafting any kind of fix. Although President Herbert Hoover had worked as an engineer before entering government, he was reluctant to intervene in the system, thinking it best to let markets sort things out themselves.

But Scott had the solution, Hubbert thought. Writing to Nell, Hubbert called Scott a "genius." Through his first months in New York City, Hubbert grew close to Scott, learning more about his theories and hearing of his globe-trotting adventures. Scott had no shortage of tales: Of growing up in Europe, where he'd moved with his father, head of the Berlin-Baghdad Railway. Of his education in Germany. Of how he'd fled Europe after the outbreak of the Great War, coming to Canada, where he ran a chemical plant, then returning to the United States to serve as technical director at Muscle Shoals.

During the Great War, Scott had been inspired by governments' success at guiding industry, ensuring that the military had enough bullets and tanks as well as rations for the troops and food on the home front. After the war, when the nation sank into a depression in 1920, Scott connected with a small band of like-minded experts, and together they pondered how the government could continue to guide industry during peacetime. Some of the experts involved formed a consulting firm called the Technical Alliance, completing a few studies for large unions like the Industrial Workers of the World and the Brotherhood of Railroad Trainmen. Scott still had an old Technical Alliance brochure, which listed the group's "temporary organizing committee," an impressive roster. One was Thorstein Veblen, an iconoclastic economist who'd written *The Theory of the Leisure Class*, which had introduced the notion of conspicuous consumption. (Ever the nonconformist and carrying on many affairs, Veblen got kicked out of one professorship after another, earning the nickname "the bad boy of economics.") Another was physicist Richard Tolman, who'd become a collaborator and friend of Albert Einstein's, helping draw out the implications of his theory of general relativity. Howard Scott, then around thirty, was one of the Technical Alliance's youngest members. On the brochure, he was listed as the group's chief engineer.

During the boom years of the 1920s, Veblen had died, and most of the other members lost interest or got caught up in other matters, so the Technical Alliance disintegrated. Scott alone had continued to focus on the goal that had brought all those experts together in the first place: how to build a new society for the machine age.

However, with the deepening Depression of the early 1930s, some of those former members showed a rekindled interest. One was Stuart Chase, the economics writer who frequented the Meeting Place. Another was Frederick Ackerman, an architect and one of America's leading experts on public housing, whose firm had designed many huge complexes in New York. Yet another was Bassett Jones, an electrical engineer who designed elevator systems for skyscrapers—including the tallest of all, the recently completed Empire State Building. Hubbert prodded Scott to round up these old members who were showing an interest. Jones and Ackerman, in particular, had free time since their work depended on the construction market, which had collapsed. They began helping Scott and Hubbert gather data and compile graphs. Together this team computed the degree of "technological unemployment"—how many people had been put out of work by machines—and measured the pulse of various industries, from iron and coal extraction to automobile and telephone manufacturing. It was just the start of what they envisioned would become a vast survey of some three thousand products, industries, and indicators.

Hubbert wanted to get Scott's ideas out to a wider audience. Despite having worked on his Theory of Energy Determinants for years, Scott had never written up a detailed description of it, nor his proposal for replacing dollars with energy certificates. Outside Greenwich Village circles and the scattered members of the defunct Technical Alliance, hardly anyone had heard of Scott.

In person, Scott was verbose—sometimes eloquent, other times overly technical, but always pouring forth a stream of concepts, historical examples, and anecdotes. Hubbert kept pushing Scott to commit his ideas to paper. The efforts led nowhere, so Hubbert began to write a book himself.

. . .

HUBBERT, THEN TWENTY-SEVEN, weighed what to do with his career. After a semester at Columbia, he was disappointed. He had thought he'd been hired to teach geophysics in a comprehensive way, training students to be equally versed in physics as in geology, and regularly employing math. Instead he found the head of the geology department to be an autocrat who demanded he teach the subject with a wholly practical slant, such as how to operate magnetometers and torsion balances. His colleagues in the department, Hubbert thought, "didn't have any more idea than a jack rabbit of what geophysics really was about."

Meanwhile, Scott's industrial survey was gaining traction. On visiting Columbia's faculty club, Hubbert met Walter Rautenstrauch, head of the university's department of industrial engineering. It turned out that years earlier Rautenstrauch had attempted something like Scott's survey. Impressed with what their small group had already accomplished, Rautenstrauch offered the use of a large, empty drafting room, in place of Hubbert's personal office.

Also in the spring of 1931 the Illinois State Geological Survey approached Hubbert, since its chief had identified him as one of the country's best-trained geophysicists. Hubbert assessed the state's plan for surveying its natural resources and proposed a promising new approach: using measurements of electrical resistivity, or how easily electricity flowed through the ground. This technique could uncover groundwater aquifers as well as deposits of fluorspar, a mineral widely used in the steel industry. Impressed, the survey's chief offered Hubbert a job.

Hubbert turned down the offer, however, since he still aimed to turn things around at Columbia. He had a few students under him and hoped to lure a few of Chicago's top geology students to Columbia's lower-ranked program. "If I get these fellows there will be two departments that will sit up and take notice," he told Nell. "Very frankly I shall consider that I have won a great victory." With his summers free, however, Hubbert arranged to spend that time in Chicago, helping the Geological Survey assess the state's resources. He could certainly use the extra money that the work would bring in.

Other than his work with Scott, Hubbert had little reason to stay in New York for the summer. During his first year there, he'd met a few women he found intriguing, but most of them were already married. In any case, he was wary of getting hitched himself. After filling Nell in on his love life, or lack thereof, he added:

> It seems to me to be quite impossible to be really free to partake
> of dangerous issues (for which I have some slight propensity) when
> one has an attached family who might become hungry as a result of
> his own personal courage. I want to be free enough that if I feel my
> individuality being cramped by Columbia University, I can walk
> out cold and have it concern no one but myself.

As Hubbert's summer in Chicago was coming to a close, his friend Yelena Pavlinova ran into trouble. She was originally from Russia, which she'd fled more than a decade earlier, during the Bolshevik Revolution. She had moved to Turkey, then finally settled in the United States. In the meantime, her region of Russia had gained independence, becoming Latvia. In the summer of 1931 the US government belatedly ruled that she had entered the country on an invalid passport—and that meant she couldn't apply for American citizenship. But she couldn't be deported, either, because neither Russia nor Latvia wanted to give her a passport. She was officially stateless. Her plight got picked up in the media and was reported across the nation. Some of the headlines sounded like want ads: "Woman Without Country Seeking Citizenship."

A couple of years earlier Yelena had chided King that marriage was not to be taken lightly. But Hubbert had spent the past year around Greenwich Village, the center of America's free-love movement. In his social circle, many saw marriage as something fluid, even outdated. Howard Scott had already been married and divorced, and at the time had a girlfriend, Eleanor Steele, who was married to another man. One of Eleanor's best friends, the anthropologist Margaret Mead, was married—

and was meanwhile having an affair with another anthropologist, Ruth Benedict. When two journalists held their wedding party at Lee Chumley's speakeasy, downstairs from the Meeting Place, one of their friends remarked, "But why did you have to get married? Isn't that a little quaint of you?"

Given Yelena's situation, she and King got married on October 3, 1931, allowing her to remain in the United States and continue her graduate research in parasitology. Soon afterward Hubbert returned to New York.

On arriving back home, Hubbert found Howard Scott in financial trouble. Scott hadn't held a regular job in years and was behind on rent for his room in a run-down Greenwich Village brownstone shared with a host of other intellectuals. Hubbert paid off Scott's back rent and got an apartment for the two of them in Chelsea, a district of tenements and warehouses north of the Village. King, Howard, and Eleanor began hosting Sunday brunches at their apartment, which became known in the neighborhood for heady discussions that stretched long into the afternoon.

The small group doing the industrial survey—Scott and Hubbert, Ackerman and Jones—was quick to gather volunteers. One who joined the effort was Leon Henderson, an economist who lived around the corner from Hubbert and Scott's apartment. Working for the Russell Sage Foundation, an organization dedicated to helping the poor and elderly, Henderson battled loan sharks. In his spare time he helped Scott and Hubbert's industrial survey by doing data analysis. Ackerman also pitched in, lining up funds from the American Institute of Architects, which paid unemployed architects and draftsmen to ferret out data from reference books and plot it on graphs. By the spring, with a staff of ten, they had titled their project The Energy Survey of North America. They also gave their organization a name: Technocracy.

As their work progressed, word got around. In February they met with Richard Walsh, head of a publishing house called the John Day Company, which was printing a series of pamphlets, including one on the Soviet command economy by Joseph Stalin. To get Technocracy's ideas out quickly and widely, Walsh wanted to publish a pamphlet by Scott.

Harper's magazine also commissioned an article from Scott—an opportunity to reach its large, liberal audience.

At the time, intellectuals and leaders had floated dozens of plans for some form of centralized planning to pull the country out of the Depression. Many on the Left favored greater control by the government—something like the Soviet Union's recent five-year plan that laid out ambitious goals for the economy, such as tripling the extraction of coal to power the nation's factories. (*New Russia's Primer*, a book describing the plan in simple terms, had become a best seller in the United States—and an inspiration to Hubbert, who was still struggling to write his own book on Technocracy.) At the opposite end of the political spectrum was Gerard Swope, president of General Electric, who proposed putting business in the driver's seat by abolishing antitrust laws and allowing corporations to cooperate in setting production levels, somewhat like in Fascist Italy. Compared with the proposals receiving attention, Technocracy's ideas—replacing dollars with energy certificates, banishing most private property and private enterprise, shoving politicians aside in favor of technicians—were far more radical. But as yet few had heard of their small group.

Technocracy made its public debut with a talk by Scott at New York's City Club in April 1932. By then some 9 million people had lost their jobs. Scott predicted unemployment would grow worse, reaching 11 million by the end of the year—and if trends continued, 20 million or more in a couple of years' time. Despite Technocracy's promotional efforts, only one media outlet covered the talk: Richard Walsh, the publisher who'd wanted to publish a pamphlet by Scott, wrote an editorial in the magazine *Judge* (the main competitor of *The New Yorker*). His piece, "The Engineers' Revolution," celebrated Technocracy's promise of abundance and security, leisure and equality, all achieved through technological progress and control by technicians. "It's the newest thing on earth," Walsh concluded. "It is hard-boiled, and it is offered on a take-it-or-leave-it basis. And it sounds to us like the only revolution worth talking about."

Hubbert certainly saw Technocracy—and his own efforts—as revolutionary. "Perhaps you would be interested to know that you are receiving a letter from one of the most dangerous radicals in the United States," he wrote to his sister Nell. "I am for an overhaul of this country from top to bottom so thorough that the plans of the Socialists and Communists look cheap by comparison." When regaling Nell with tales of his exploits, he'd often exaggerate. This time was different. "This all may be written in a vein of apparent foolishness but I am really serious," he told Nell. "Unless I get shot or locked up you will hear more of this as time goes on and you will probably read it in the newspapers and magazines."

6

The Fad

IN THE SUMMER OF 1932, Hubbert went off to work in Illinois again, and on returning to New York in the fall, he found Technocracy in a frenetic rush. Aided by more grants, it had boosted its staff to thirty-six, about triple their size in the spring. Over the summer, a talk of Scott's had received coverage in the nation's most prestigious newspaper, *The New York Times*. Then a Columbia press officer had stumbled across the group and publicized it, leading to another *Times* article. By the fall, the media attention snowballed, with liberal stalwart *The Nation* declaring Technocracy's work "the first step toward a genuine revolutionary philosophy for America." Soon afterward *Time* magazine ran a piece, "Technocrats," that quoted Scott: "We are now in a dynamic system compared with the static system of history. The controls of the static age, namely the price system of production, are opposed to the controls which must govern the dynamic age of Technology."

With this growing attention, Technocracy became swamped with letters from across the country and with visitors who flocked to its offices. Some offered help, others requested statistics, and still others shared heart-wrenching stories of hardship and begged for work. Meanwhile Hubbert's efforts to write a book on Technocracy were stalled. "What I wanted to do was to get on to the technical writing," he later recalled. However, "through the emergency of the situation, the demand of the public to have something to do, we had to try to get some kind of an organization operating." By November, Technocracy received still more funds—this time from a New York City program, the Emergency Unemployment Relief Committee—and hired dozens more, swelling its staff to one hundred employees. The organization opened more offices, one near Greenwich Village and another on Park Avenue in the same building as Bassett Jones's engineering firm. Technocracy's staff became such a common presence in Columbia's engineering and business libraries that the university's annual report noted the Technocrats had "a general avidity for statistics that won the sincere respect of those in charge of the rooms."

MEANWHILE THE ECONOMY continued to crumble. By the time of the presidential election in November 1932, unemployment had passed 11 million, more than 20 percent of the workforce. It was as bad as Scott had predicted in his City Club talk in the spring. Technocracy expected that the nation would find it difficult to restore employment to earlier levels. According to its calculations, even if the economy recovered to its level before the crash, because of ongoing improvements in mechanization, only half those who'd lost their jobs would be able to find work again.

The Democratic candidate for president—Franklin Roosevelt, then governor of New York—was optimistic about America's ability to find its way out of the Depression. Nonetheless, Roosevelt's longer-term outlook had some similarities with Technocracy's. "It seems to me probable that our physical economic plant will not expand in the future at the same rate at which it has expanded in the past," he said in a speech on the campaign trail. "We may build more factories, but the fact remains that we have enough now to supply all of our domestic needs, and more, if they are used." In another speech, he declared that "our last frontier has long since been reached, and there is practically no more free land." The nation had reached a mature stage, in which it was time for a "reappraisal of values," turning from an emphasis on growth to the "soberer, less dramatic business of administering resources and plants already in hand," and "of distributing wealth and products more equitably, of adapting existing economic organizations to the service of the people."

In rebuttal, President Hoover declared, "I do challenge the whole idea that we have ended the advance of America, that this country has reached the zenith of its power and the height of its development." He added, "If it is true, every American must abandon the road of countless progress and countless hopes and unlimited opportunity."

Voters were apparently disillusioned with Hoover's optimistic statements that failed to bring results. On November 8, 1932, Roosevelt defeated Hoover in a landslide, 472 electoral votes to 59.

WITH THE CAMPAIGN over and four months to wait until Roosevelt's inauguration in March, the press glommed onto Technocracy, giving it even more coverage than Roosevelt's plan for reform, the New Deal. The surge of interest sometimes came from unexpected corners. "It is not necessary to take all our engineers' equations and curves as an exact mathematical guide to future events to realize that they contain a fearsome suggestion," argued a *Wall Street Journal* column by Thomas Woodlock, the paper's former editor-in-chief. "One thing is certain," Woodlock concluded. " 'Technocracy' is going to become a much more familiar word than it is at the moment." To handle the growing attention, a couple of public relations men, Charles Bonner and Roger William Riis, volunteered their services and created a new venture, the Continental Committee on Technocracy. As its head, they recruited Langdon Post, a former New York City Council member and a friend of Roosevelt's.

Many commentators got the impression Technocracy blamed technological progress for the nation's problems. "Is the machine the 'ogre' that has caused the economic distress of the last three years?" asked one newspaper article about Technocracy. A long piece in the *Los Angeles Times* was titled simply "Will a Machine Get Your Job?" The Technocrats tried to explain that they did not condemn the machine and actually celebrated it. The real problem, they argued, was the system of distribution—the price system.

As Technocracy gained more attention, various business leaders and experts spoke out in opposition to its ideas. "They are wrong," declared engineer Charles Kettering, head of research at General Motors. "They overlook the fact that man hates monotony." The Technocrats had proposed building cars that would last fifty years. Kettering countered, "This automobile would not be worth anything except to a junkman in ten years, because of the changes in men's tastes and ideas. The desire for change is an inherent quality in human nature."

Even with all the discussion of Technocracy, many were still flummoxed by its technical jargon and abstractions like the price system. "This Technocracy thing, we don't know if it's a disease or a theory," wrote humorist Will Rogers in his nationally syndicated column at the end of 1932. It may turn out to be a fad, he added. "But people right now are in a mood to grab at anything."

Many reporters looked to others to make sense of it. When ships arrived in US ports and reporters boarded to interview arriving celebrities, their first question was often: "What do you think of Technocracy?" One such celebrity, rumored to have an interest in the movement, was the world's most famous scientist, Albert Einstein. In early January the wild-haired physicist arrived in Los Angeles from Germany and was asked the standard question. Einstein replied, "Technocracy? *Was ist das?*" (What's that?)

There to pick up Einstein was his friend Richard Tolman, the physicist who'd had some involvement with the Technical Alliance a decade earlier. Tolman explained that Technocracy's aim was to have engineers and scientists direct the economy. "*Ja*," Einstein said. "The problem of getting the men who understand most to take charge of government is the most difficult problem in the world. It always has been. It is not yet satisfactorily solved." (A few weeks later Adolf Hitler would take charge in Germany.) Soon afterward Einstein gave a speech describing the roots of the Depression in terms similar to the Technocrats', explaining that improvements in machines had "decreased the need for human labor" and "thereby caused a progressive decrease in the purchasing power of the consumers."

Another celebrity of sorts weighed in from Rome. "Technocracy Will Not Work, Benito Mussolini Declares," read one US newspaper headline. The article, written by the Italian premier and founder of the Fascist movement, argued that rather than rule by a specialized group—whether soldiers, theocrats, or technocrats—Italy and the United States alike needed rule by a "lone guiding mind." A dictator.

Amid this "technocraze," as many began calling it, a reporter at the conservative *New York Herald-Tribune* grew skeptical. Swirling around Scott were too many fantastical stories: that he'd been engaged in engineering projects around the world, in a wide variety of fields—railroads and electricity transmission, munitions and nitrates; that his family had a fortune in Constantinople that had been seized during the Great War; that he'd built Canada's only plant for manufacturing the chemical acetone; that he'd turned down professorships to continue working on his Theory of Energy Determinants. All the jobs, travels, and inventions seemed too much for a man of just forty-two to have accomplished—especially since many around Greenwich Village knew that for the previous dozen years, Scott had spent nearly all his time there, hanging out in teahouses and clubs like the Meeting Place.

The reporter, Allen Raymond, dug in. He tracked down the actual chief of the Berlin-Baghdad Railway—who wasn't Scott's father and had never heard of Scott's father, either. Raymond discovered that, at the Muscle Shoals nitrate plant, Scott had not been chief technician. He'd actually been an equipment superintendent with the task of redesigning the factory's furnaces. When he'd made derisive remarks about American and British efforts in the war, his colleagues became suspicious, and his boss had then belatedly vetted Scott's CV and found that he apparently never held a job that he had stated on his application form. Under pressure, Scott had left Muscle Shoals after less than six months.

During an interview with Raymond, Walter Rautenstrauch referred to Technocracy's leader as "Dr. Scott." Raymond asked why he used that title. Because, Rautenstrauch said, Scott had a doctorate from the Technische Hochschule in Berlin, one of Europe's top technical universities.

Raymond had already checked into it. He informed Rautenstrauch that the university denied having ever awarded Scott a degree.

"But he told me so himself," Rautenstrauch replied. "And King Hubbert certified it to me."

Raymond then asked Hubbert, who said he didn't know whether Scott had a degree but did attest: "Scott has the greatest amount of knowledge of the physical sciences of anyone I have ever met. I would testify anywhere as to his scientific attainments."

When Raymond asked the head Technocrat himself, Scott admitted he had no doctorate. He had no degrees at all.

Raymond wrote a series of articles in the *New York Herald-Tribune* that got republished across the country, from the *Springfield Republican* in Massachusetts to Portland's *Oregonian*. "The real Scott," Raymond wrote, "was a man quite different from the Scott who was being portrayed to the public by associates in Technocracy."

BY THE END OF 1932, the Depression had entered its fourth year. "There is not a man in the whole world today that people feel like actually knows what's the matter," Will Rogers charged. "Our 'big men' won't admit they don't know. They just keep on hoping they can bull their way through."

So despite Raymond's exposé, the interest in Technocracy continued unabated. "Everybody from bank president to panhandler is talking these days about 'technocracy,'" reported the *Chicago Tribune* at the close of 1932. "The amazing thing is the intense interest. It is the favorite topic of discussion at the Chicago Stock Exchange. The elevator boy asks you what it is. It is talked about over the bridge table and in the bread line." When *Illustrated Daily News* in Los Angeles ran a series of front-page articles on the topic, "At first we were thought crazy, but our first edition was a sell out," said Manchester Boddy, the newspaper's publisher. "It has offered our first opportunity to give the reader a rational explanation of the depression. Naturally they eat it up."

When Scott gave his highest-profile talk yet—at the Hotel Pierre on Fifth Avenue, one of New York's elite locales—it attracted intense interest and was broadcast over a nationwide radio hook-up. However, he rambled incoherently through much of his speech. "We are not attempting to say, as some of our critics have said, that there is going to be chaos or there is going to be doom," he said. He then proceeded to issue an ominous forecast: "Unemployment in the United States in eighteen months, if present trends continue, will exceed twenty million"—nearly double the number of jobless at the time. Unless the nation made drastic changes, he predicted, "then I am afraid we are in for the gravest social trouble that this country has ever experienced."

After this speech, many of those who'd supported Technocratic ideas were frustrated. "The whole country was leaning forward eagerly to catch every word uttered by Mr. Scott," wrote Harry Elmer Barnes, a sympathetic columnist at the *New York World-Telegram*. But Scott's "unnecessarily incompetent and insolent performance" let down "both the curious and the converted." Scott blamed his poor performance on a bout of the flu. Others in the group charged that he'd abandoned a carefully crafted speech and instead tried to wing it.

Soon after the Hotel Pierre disaster, many of the core Technocrats assembled to talk things out. Dissenters brought up a key issue: would the organization be democratic, taking votes on how to proceed?

Scott didn't answer, but his girlfriend, Eleanor Steele, spoke up on his behalf. "Of course it will be democratic," she said, "but Howard should always have the power of veto."

Scott remained silent.

"That decided it so far as we were concerned," recalled Bonner, the public relations man who'd helped establish the Continental Committee on Technocracy. He felt that Scott wanted to be dictator—or, at least, to be free to do whatever he pleased. Bonner and many others were ready to split.

Even Scott's longtime associates, Bassett Jones and Frederick Ackerman, were disgruntled, as were two of Technocracy's early supporters, Leon Henderson and Walter Rautenstrauch. The four agreed to split from

Scott's group, announcing their departure in the *New York Evening Post*. Columbia also severed ties with Technocracy, declaring that the organization was no longer welcome on campus and that its staff and funding would remain under the university's control. This repudiation made the front pages coast to coast. On January 24, *The New York Times* said, "Scott Is Ousted from Technocracy by Split in Group," and in California the *Oakland Tribune* announced, "Rebels Oust Technocracy High Priest."

It had been only a week and a half since Scott's Hotel Pierre talk, and suddenly the group had lost almost everything. Scott retreated to Technocracy's office on Park Avenue, but reporters persuaded him to hold an impromptu press conference. Photos printed in newspapers across the nation showed Scott towering, flanked by Technocracy's two core members who remained loyal: Dal Hitchcock, an analyst and writer, and King Hubbert.

Even after this debacle, Technocracy attracted intense interest. "Everybody I talk to these days wants me to discuss Technocracy," said automobile magnate Henry Ford. He partially disagreed with Technocracy's diagnosis of the economic crash—but like Technocracy, he held a utopian vision of the future. "In the real machine age which is to come," Ford said, "the dirt and ugliness and confusion and noise and disregard of human rights which are all about us today will be done away with."

ON MARCH 4 came Franklin Delano Roosevelt's long-awaited inauguration. Troops camped outside major cities—in case, the secretary of war said, of action by "Reds and possible Communists."

His voice resonant and his strong jaw thrust out, Roosevelt opened his speech: "Let me assert my firm belief that the only thing we have to fear is fear itself." He inspired confidence in a way that Hoover hadn't for years, if ever. Thunderous applause erupted from the crowd. "Plenty is at our doorstep, but a generous use of it languishes in the very sight of the supply," Roosevelt continued. "Primarily this is because rulers of the exchange of mankind's goods have failed through their own stubbornness and their own incompetence, have admitted their failure and abdicated."

With the banking system teetering on the edge of collapse, on Roosevelt's first day in office, his administration immediately closed all banks, nationwide. Then they evaluated each bank, allowing only those that were sound to reopen, which managed to halt an epidemic of bank failures. By the summer, when Roosevelt completed the first hundred days of his presidency, his administration had requested all manner of bills—for agricultural reform and unemployment relief, for slashing salaries of federal workers and veterans' benefits, for dictatorial power over currency and foreign exchange. The vast majority of these proposals sailed through Congress.

With Roosevelt's New Deal taking such steps—far bolder than anything Hoover had attempted—the spotlight turned away from Technocracy. Hubbert finally had some breathing space, sending his sister Nell his first update in several months. "So much hell has broken loose," he told her, which left him exhausted. "I've been living a double life," he joked, "by being a harmless college professor by day and a Terrible Technocrat by night."

Despite the flurry of bold action in the nation's capital, Hubbert felt the situation was dire. "You had better sit tight for this winter," he warned Nell. He expected it would be worse than the winter they'd just been through, when more than a quarter of the workforce was unemployed. "Keep a big supply of food on hand," he advised, "just in case money should become either worthless or non-existent."

7

Lessons

IN A HANDFUL OF CITIES across the country—from Washington, DC, to Los Angeles—fans of Technocracy had formed study groups to learn more about the group's ideas. As yet, Technocracy had issued little of its own material, and others were rushing in, publishing pamphlets and books and even establishing their own publications, such as *Technocracy: The Magazine of the New Deal*. To try to control the use of their name, the Technocrats filed incorporation papers, so the group became Technocracy Incorporated.

Then in the fall of 1933, Hubbert knuckled down to finally write his book. He had moved out of his apartment with Howard Scott and gotten a small place of his own. He outfitted it with new furniture, including a daybed with a pillow that featured Technocracy's new logo, a yin-yang symbol in red and gray, which the group called by its old European name, the "monad." (To them, it represented balance: running society efficiently and consistently, avoiding booms and busts.)

Hubbert aimed to write a "comprehensive treatise," he told Ralph Chaplin, a veteran labor activist who was helping Technocracy organize in Chicago. "I am going to try to ditch everything in favor of that end." His goal was to write something "irrefutable and loaded with a wallop that won't be readily forgotten." He explained, "I want to put the thing in such a form that it will not be possible to attack us without first demolishing the outstanding concepts of science itself. In other words I want to make all of our conclusions follow directly as the logical and necessary consequences of fundamental science."

In December, Hubbert came to Chicago to attend the Geological Society of America's annual meeting, and while in town he lectured to a Technocracy study group there. But the writing was still proving difficult. As he wrote to Nell after Christmas break, "I've got to write a book this spring or bust."

In March 1934, Hubbert copyrighted the first section of his book, titled *Technocracy Study Course*. The first four chapters—or "lessons"— covered the basics of science that he felt every Technocrat should understand. The first lesson, "Matter," explained how all things are built of molecules and atoms. Later lessons covered energy and the laws of thermodynamics, explaining how energy is always conserved, never destroyed. In April the first mimeographed versions of these lessons went out to the far-flung study groups, by then established in more cities, from Washington, DC, to Denver, Colorado.

That summer Hubbert's usual work with the Illinois State Geological Survey was canceled since the state was short on funds. But at the federal level, the New Deal was still picking up speed, directing spending to hundreds of projects, including research. The US Geological Survey was flush, so it hired Hubbert to continue his resistivity surveys, work that took him beyond Illinois into Kentucky.

Hubbert was able to hire two crew members, the kind of position that usually went to junior men, still in college. But "this was in the deep dark days of the Depression," Hubbert recalled, and he received many applications from seasoned researchers desperate for work. "It was kind of heartbreaking to read these lists of applicants." He could afford to be choosy and sought people with solid knowledge of physics and math. From graduate geology courses in Chicago, he remembered a friend, Darrell Hughes, who'd gone on to postdoctoral work at the California Institute of Technology but then had not been able to find regular work. Hubbert offered Hughes a job, admitting the pay was "an insult," but Hughes took it.

Over the summer, Hubbert continued writing more lessons in the study course. He criticized mainstream thought about the long-term outlook for the future—in particular, for economic growth. Up until the Depression, economic growth had been running about 5 percent a year—and many had come to think of this rate as normal and that it could continue indefinitely, following growth curves that arced upward with no end in sight. "It has come to be naively expected by our business men and their apologists, the economists, that such a rate of growth was somehow inherent in the industrial processes," Hubbert wrote.

However, since Earth is finite, it was physically impossible for that kind of growth to continue indefinitely. Instead industrial activity would more likely follow a different type of path, Hubbert maintained. There was only so much land, so much coal, so much iron. At some point growth had to hit limits—as shown in an elegant experiment by the biologist Raymond Pearl. He put a breeding pair of fruit flies in a bottle along with some food. From this initial pair, the population inside soared. But the growth later tapered off, and eventually the population reached a plateau. A graph of the population over time formed a so-called S-curve—looking like a stretched-out S, it reflected the phases of rapid growth, slower growth, then no growth at all.

Like the fruit flies, Hubbert argued, many facets of industrial society would also follow an S-shaped growth curve. There were only so many rivers to be dammed, for example. If hydroelectric power were pushed to its utmost, it would similarly go through a phase of rapid growth, then slower growth, finally reaching a plateau, producing a constant amount of electricity—a condition that could last indefinitely.

In addition to this S-shaped curve, Hubbert showed another type of growth curve, with a rise, peak, decline—and finally a stabilization at some lower level. This could represent forests that were getting logged out, where the rate of logging couldn't continue increasing. The annual timber harvest might drop, then settle down at a lower rate that could be sustained for the long term.

Finally, he showed a curve for nonrenewable resources, such as

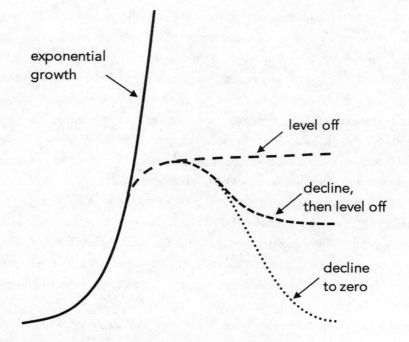

Growth curves: Hubbert's *Technocracy Study Course* showed curves representing four types of growth.

metals and fossil fuels. Their extraction could soar, but not forever. A perfect example was the "life history of a single oil pool." Hubbert explained:

> The production rises as more and more wells are drilled, until it reaches a peak. From that time on the production declines year by year, until finally it becomes so small that the pool is abandoned. In most American oil pools the greater part of this history takes place within 5 to 8 years after the discovery, though the pool may continue to be operated for the small remaining amount of oil for 10 or 15 years longer.

Hubbert described the frantic drilling as "the mad business rush of big and little oil companies, like so many buzzards fighting over a carcass, each trying to get his share, while the pool, in the meantime, is being drilled as full of holes as a pin-cushion."

After laying out these four types of growth curves, Hubbert reviewed the histories of various industries—the same type of data Technocracy had compiled for its Energy Survey of North America. Iron production, railroad freight, automobile manufacturing, the total production of energy—all had grown, then leveled off. "The persistent S-shape of each of the growth curves examined is a striking and singular phenomenon," he wrote. "In the United States that period of most rapid growth has passed, and already more or less unconsciously we have entered well into the second period of growth, that of leveling off and maturation."

By the time Hubbert copyrighted Lessons 5 to 16 of the *Study Course*, in July 1934, Technocracy had gained more members and grown increasingly organized. They had set up a system in which anyone could start a new "section" in their city, as long as they could gather twenty-five members, each willing to pay annual dues of five dollars. Some sections formed their own version of Boy Scouts and Girl Scouts—called "Farads," after a unit of measurement for electrical charge.

With a large portion of the *Study Course* already distributed to Technocracy's sections across the country, Hubbert wrote a memo regarding the status of the study groups, addressed to "All Officers and Members of Technocracy, Inc." He sounded like a frustrated instructor—and he was. "The rapidity with which the various groups have covered the first sixteen lessons," he wrote, "raises a question as to the manner in which classes have heretofore been conducted." The *Study Course*, he explained, was meant to serve as only a bare outline of a wide variety of topics, and he expected the study groups to read further, drawing on the work cited in his book as well as doing exercises, such as collecting data and plotting it on graphs and making maps of world resources. Such study would raise important questions, Hubbert wrote, such as "Where does Japan obtain the oil she requires to operate her battle fleet? What would happen if she were blockaded?"

During his break at the end of 1934, Hubbert went to Rochester, New York, to stay with a friend. A reporter tracked him down and asked about rumors that Technocracy was dead. "Two years ago there were only fifteen of us," Hubbert responded. "Now we have offices from Miami to Edmonton and from New York to the Pacific Coast."

. . .

ALL THROUGH HIS time in New York, with his teaching and his Technocratic work, Hubbert had found little time to look for love. In any case, he'd always been hesitant to settle down. But over the summer, while in Kentucky, he'd gotten to know some local geologists who invited him to dinner a few times. Apparently noting the lack of a ring on his finger, and hearing that he lived in New York City, one of them had suggested Hubbert meet their former classmate who lived in Manhattan, working as a medical secretary for a large pharmaceutical firm. Her name was Miriam Berry.

Soon after returning to New York, King Hubbert got to meet Miriam, who was brilliant and science-minded, in her late twenties, just a couple of years younger than he. Her job went far beyond that of a typical secretary. She read the scientific literature and corresponded with doctors on technical matters. She and King had a common interest in medicine. He had learned a fair amount about the subject back in Chicago, from his roommates in the boardinghouse. "I saw those boys through medical school and internship," he recalled, joking, "I'm practically a doctor by proxy."

Miriam had separated from her husband but was still married. King remained married to Yelena, so she could stay in the United States. Nonetheless King and Miriam quickly became an item.

Hubbert also began to make more connections with other researchers. "I didn't have very much intellectual companionship" at Columbia, he recalled. He craved more interaction with geologists—in particular, those who were well versed in physics. Through his summer work, he'd gotten to know some of the researchers at the US Geological Survey and was impressed with them. They were apparently intrigued by him as well—including his wider ideas on society. In early 1935 the Geological Survey invited Hubbert to its DC headquarters to give a talk explaining Technocracy's ideas to its top men. He covered Technocracy's notion of "continental control": that all of North America, from Canada to Venezuela, should be united under one government, which they called the "Technate," to enable the continent to run efficiently and strengthen its defense. Technocracy's vision was like that of Manifest Destiny of decades past. "We had to have it," Hubbert argued. "It was our sphere of control." However, the survey men didn't buy it. "It was a good honest, frank discussion and we all disagreed and had a wonderful time," Hubbert recalled several years later. "I was regarded as being absolutely crazy."

Despite such clashes, he bonded with some of the men at the survey—in particular William Rubey, a geologist five years older. Rubey was kind and considerate, modest and tactful—whereas Hubbert was rarely any of these. In science, though, they shared an aptitude for physics and a drive to apply it to understanding the Earth. Soon after returning from his trip to DC, Hubbert wrote Rubey a long note. "The whole visit at the Survey was to me like a breath of fresh air," he said, since "intellectual sterility is the rule rather than the exception around Columbia." He added, "I need something like that ever so often to serve as a gauge in order to make sure that I am not going nuts, or at least not entirely alone in my particular brand of insanity."

BY MID-1935, AFTER Roosevelt had been in the White House for a couple of years, his New Deal measures were having a tangible effect. Instead of sinking deeper into the Depression, the US economy showed moderate growth, and unemployment was gradually shrinking. Despite these successes, Hubbert thought the New Deal was a last-ditch attempt to maintain economic growth without making required fundamental changes. He thought it wouldn't work. "The period of industrial expansion under Price System dominance is virtually over," he wrote in further lessons in the *Technocracy Study Course*, laying out the group's vision for the future. Speaking to Technocracy's section in Cleveland, he forecast, "The American Price System in all probability will have reached a crisis before the next Administration period is over"—that is, by 1940.

The problem, as Hubbert saw it, was that the New Deal had been using deficit spending to prop up the economy, building up ever-larger debt. But if this spending stopped, Hubbert predicted in *Technocracy* magazine in December 1935, it "would in short order shut down the country tighter than it has ever been shut down before."

There was, Hubbert argued, one way out of the predicament: "a nice friendly war with somebody." He added sarcastically:

> Consider the advantages: there would be an industrial boom, turning out munitions. By one move the unemployment and relief problems would be solved by putting the abler members in the army (and hoping they would get shot), and the remainder to work in the war industries. War profits would follow, new fields for investment would open up, and even the interest rate might be jacked up a little. And a big time would be had by all—who lived through it in one piece.

Even if war could be a way out of the country's economic woes, Technocracy opposed this path. The organization imagined that all of North America could become one Technate, governed according to the system that Technocracy laid out, and that this would be achieved by popular demand and without force. Yet Hubbert saw little hope for reform through conventional politics. Propaganda techniques, developed during the Great War, had been put in service of advertising and elections—and people purchased and voted accordingly. Meanwhile all the real decisions got made behind closed doors. As he wrote in the *Study Course*, "No question of really fundamental importance is ever submitted to popular election."

Hubbert was harshly critical of the irrationality and waste, the corruption and inequality, of the current system. Yet he didn't blame anyone. Given a particular situation, most everyone would react the same way, he argued. The solution lay neither in moralistic sermons nor in better legislation. Only a complete overhaul—of the government, of money, of industry—would truly change the situation.

Hubbert concluded the *Study Course* with a final lesson, "Industrial Design and Operating Characteristics." It described how the Technate would produce optimal products, balancing function and durability against the required energy inputs. The Technate's cars would be something between the everyman's Chevy and the luxurious Delaunay-

Belleville, and would run on diesel, providing higher mileage than gasoline engines. Rather than each family owning its own car, people would rent them as needed from a network of garages, using energy certificates to pay for the mileage driven.

The Technate would likewise optimize communications, possibly replacing all mail with signals sent by wire. Housing would be modular, with thick insulation to keep heating and cooling costs to a minimum. "New towns and cities would have to be designed as operating units from the ground up," Hubbert expected. But this revamped society would be so efficient that "all this will be accomplished with a shortening, rather than with a lengthening of the working day."

By the time Hubbert finished writing the book, Technocracy had become more formalized. It had instituted "regulation dress": for the men, a gray business suit with blue tie and matching handkerchief, and for the women (about half of Technocracy's members) a similar suit but with a skirt. It also issued small lapel pins with the monad symbol—the crimson and silver yin-yang. Hubbert wore his lapel pin faithfully, even at Columbia, where Technocracy was unwelcome.

The Technocrats had many of the details of a new system sketched out. However, despite years of working with Scott, Hubbert still had little to say about how to achieve the organization's goals.

Borderlands

EVEN IF MANY OF HUBBERT'S fellow geologists regarded his Technocratic ideas as screwy, they respected his scientific work. Although he had yet to earn his doctorate, in 1935 the American Geophysical Union had chosen him as one of a few experts in geophysics to serve on a committee concerning "continental evolution," exploring questions such as how mountains formed. In 1936 he received another such invitation, this time from the National Research Council, an offshoot of the National Academy of Sciences, an august coalition of scientists who advised the federal government. They chose Hubbert to join the Committee on Borderlands—short for the formal name, the Interdivisional Committee on Borderland Fields Between Geology, Chemistry, and Physics.

Hubbert had been trying, ever since joining Columbia, to knock down the walls between these fields. This committee didn't go as far as Hubbert wanted to, but it was operating in a similar spirit. In December 1936 he traveled to Washington to meet the other nine members and discuss how to promote interdisciplinary research. They drew up a wish list for future projects, and each committee member volunteered to probe an item on the list. One of the questions: Can we develop a theory to make scale models that are actually realistic?

Hubbert had been following work with such scale models for the past decade, ever since early in graduate school, when he had built his own model, using clay packed into a wooden frame to represent a fault. Geologists had employed scale models for more than a century, going back at least to the early 1800s. Their aim was to simulate features of the earth such as mountains and river basins, to understand how they may have formed. But the models often had little resemblance to the real thing and gave varied results, so it was difficult to argue that one model was better than another.

When the Borderlands Committee called for someone to work out a theory to guide scale models, Hubbert spoke up: "I'll take that."

WELL BEFORE HE knew a name for it, Hubbert had been intrigued by scaling. He'd noted how things of widely varying sizes could behave in strikingly different ways. As a child, he'd seen a mouse jump off a table and scurry away unharmed, and he wondered how that was possible. If he jumped from a comparable height, relative to his size—say, the top of a tall tree—he'd break his legs. On the farm, he'd watched big windmills turn lazily in a breeze, while in the same breeze his toy windmill spun furiously.

During his time in Chicago, Hubbert had picked up a partial answer to such questions from his mentor Bretz. To support their weight, Bretz had explained, large dinosaurs needed thick bones—much thicker than if you simply scaled up a person or a dog severalfold. While in graduate school, Hubbert had also read Galileo's classic *Two New Sciences*, published in 1638, which discussed these same issues:

It would be impossible to build up the bony structures of men, horses, or other animals so as to hold together and perform their normal functions if these animals were to be increased enormously in height; for this increase in height can be accomplished only by employing a material which is harder and stronger than usual, or by enlarging the size of the bones, thus changing their shape until the form and appearance of the animals suggest a monstrosity. . . .

Whereas, if the size of a body be diminished, the strength of that body is not diminished in the same proportion; indeed the smaller the body the greater its relative strength. Thus a small dog could probably carry on his back two or three dogs of his own size; but I believe that a horse could not carry even one of his own size.

At Chicago, Hubbert had studied scale models with Rollin Chamberlin, who'd often used them in his own work. Various researchers had built models of mountains and valleys, faults and fractures—but the models often had unrealistic features. One well-known geologist, Bailey Willis, simulated the formation of mountains in the Appalachian Range using beeswax mixed with plaster of Paris. Willis placed this waxy plaster inside a large oak box, and a piston compressed the "rock" from one side, but it folded and lifted up out of the box. To hold the "rock" down, Willis placed five hundred pounds of lead shot on top of the wax—the equivalent of the simulated rock being buried under another geological layer a couple of hundred miles thick. Then when applying pressure from the piston, the wax folded and buckled, somewhat like in real mountains. However, the model was meant to simulate geological formations at the surface, not buried deep, so the layer of lead shot on top was utterly artificial.

In 1933, when Hubbert had attended the International Geological Congress in Washington, DC, he'd seen a large table holding a display of about a dozen scale models of mountains, each occupying about one square foot. He studied them intently, marveling at how, unlike others he'd seen and read about, each of these looked like the real thing, only in miniature, with faults and fractures in the correct proportions. The geologist who had made the models—Hans Cloos, a small, wiry German in his late forties—came over to chat.

"What material did you use?" Hubbert asked.

"Very soft, almost liquid clay," Cloos said.

"By God," Hubbert shot back, "that's even worse from reality than these other things which we use."

Most models used harder materials, such as waxes, stiff clays, and plaster. While not as hard as rock, at least these materials were solid—not soupy like Cloos's models. Yet the fidelity of Cloos's models impressed Hubbert.

Back in his hotel room, Hubbert churned through calculations to determine what kinds of materials would be best to use for scale models. He realized that when you scaled down the size of the model, you also had to scale down the strength of the material used to build it—so Cloos's soupy models made perfect sense.

WHEN HUBBERT JOINED the National Research Council's Border-lands Committee in 1936, he told the other committee members about the work he'd already done on scaling. They were enthusiastic, telling him he should definitely write a report on this topic. A few weeks later, over Christmas break, Hubbert began to lay out his ideas systematically.

The basic ideas of scaling had been around for hundreds of years, Hubbert explained. Galileo had intuited many of the crucial concepts, but it took later scientists, including Isaac Newton and others, to work out how to apply these ideas rigorously and mathematically. By the twentieth century these mathematical scaling laws were widely used in engineering to build models of buildings and bridges to test their strengths, and of airplanes to examine their aerodynamics in wind tunnels. The Army Corps of Engineers likewise built miniature versions of canals and harbors to study large earthmoving efforts.

Unfortunately most geologists had neglected such work, Hubbert argued, and built their models relying more on intuition than on mathematical analysis. Simulating large phenomena such as mountains and faults required them to be scaled down roughly a millionfold to fit on a tabletop. So if done properly, Hubbert argued, the strength of the materials would also need to be scaled down similarly, making them very soft, as in Cloos's models.

In his report, Hubbert wrote out a general theory that explained how to calculate these scaling factors. Using this theory, a researcher could specify the size of the model and what kinds of forces they would put on it—such as force from a piston pushing from one side—and then compute the strength of the materials they needed to use. Scaling relationships could also help with thought experiments, Hubbert argued. "In many cases it will allow one to bring a problem of which he has little sensory experience down to a scale of things with which he is already familiar, and thereby enable him to have some intuitive understanding of it, even if no experiment is performed." For example, he pointed out, "when bird shot from a shot-gun is fired into a bank of soft mud, an individual shot will make a hole about the size of a man's fist." This, his theory suggested, was a fairly accurate model of a large meteorite slamming into Earth.

Doing his own thought experiment, Hubbert realized that his approach resolved a long-standing paradox in geology. As an undergraduate, on his first geology field trip, he had seen large rock formations folded as if they were made out of putty. And yet on a small scale, the rocks were extremely tough—after all, the quartzite often ripped the hobnails out of their boots.

Later in his education, he'd learned how seismographs could detect faraway earthquakes, picking up shock waves after they had passed through deep layers of the planet. These measurements showed that after a sharp shock, Earth reverberated as though hard as steel. On the other hand, other studies showed the continents were also soft, having deformed under the weight of ice sheets during past ice ages. Such pieces of evidence had accumulated that supported two seemingly contradictory theories, nicknamed the "hard rock" and "soft earth" theories. Geologists largely split into two camps, never able to resolve which theory was correct.

Hubbert argued his new approach resolved this conundrum. By rigorously examining the length scales and time scales involved, "the paradox of an earth, apparently both strong and weak, vanishes completely," he wrote. "We see that strength and weakness are purely relative terms devoid of meaning unless the size of the body is specified."

This paper on scale models was his first major geology publication since he'd come to Columbia. He'd finally hit upon a big problem and cracked it. In addition to sending his report to the National Research Council, he submitted it to the prestigious *Bulletin of the Geological Soci-*

ety of America. The journal soon accepted it, running the sixty-one-page paper in the October 1937 issue under the title "Theory of Scale Models as Applied to the Study of Geologic Structures."

Within months Hubbert's paper made "quite a ripple in the geological profession," he wrote to Nell. The ideas were quickly accepted, with no major rebuttals, even though he critiqued work by several top geologists, arguing they had gone wrong in building their models. Although his paper hadn't highlighted any practical applications of the ideas, Gulf Oil, one of the nation's larger petroleum companies, sent a man to New York to pick Hubbert's brain.

Chamberlin liked Hubbert's paper and had wanted to publish it in his *Journal of Geology*. But the Geological Society of America had replied to Hubbert first, so he'd gone with them. Nonetheless, Chamberlin recognized Hubbert had done a significant piece of work. Though Hubbert had been away from Chicago for years, the university accepted his paper as a thesis, awarding him a doctorate.

The members of the Borderlands Committee had been treating Hubbert as though he'd already earned his Ph.D., referring to him as "Dr. Hubbert." Finally, officially, he'd earned that title.

The Swiftest Decline

"I HAVE ONLY NOW GOT to a position where I can afford to fight," Hubbert wrote to his sister Nell, "so the war is on!" Hubbert had decided to take on Columbia as well as the wider geological community. After receiving his doctorate, in early 1938 he presented a memo he'd written, laying out his ideal for geophysics education. He demanded the university allow him to set up a comprehensive geophysics program and build a lab and workshop outfitted with ten thousand dollars worth of equipment—or else he would quit. "Otherwise," he thought, "it would be suicide for me to remain there."

His gambit didn't work. He told Nell of his clash with Columbia, explaining, "We have politely told each other to go to hell so I'll be getting out sometime in the next year or so."

It was not a good time to look for a new job.

BACK IN EARLY 1937, when Roosevelt began a second term as president, the future had looked bright. Many New Dealers declared that the Great Depression was over. Almost every economic indicator was heading upward. Government, media, and the public alike were optimistic that conditions would continue to improve. Despite this success, Roosevelt was under fire from a coalition of senators who penned a "Conservative Manifesto" that accused his administration of overspending. Trying to quell these attacks, Roosevelt announced that, for the first time in his presidency, the federal government would cut its spending to balance the national budget.

However, some in the government were uneasy about this—including Leon Henderson, who had helped Technocracy in its early days. Henderson had since moved into the federal government, eventually joining the biggest New Deal agency, the Works Progress Administration, serving as one of Roosevelt's top economic advisers. In a March 1937 memo, Henderson aired his concerns about the economy, arguing the recent recovery was all too much like the time preceding the 1929 crash. There was still a chasm separating rich and poor. Consumer spending was up, partly because earnings were rising—but also in large part because people were borrowing to buy things they couldn't afford.

Henderson's warning was fairly tepid, however, and Roosevelt went ahead with his plan to balance the budget. By that time, industrial activity had almost climbed back to pre-Depression levels. But after the budget cuts, the economy took a nosedive, and "the major part of this achievement was lost in a few months," another of Roosevelt's top economic advisers wrote in an internal memo. By November 1937, *Time* magazine declared, the plummet was "the swiftest decline in the history of US business and finance."

A couple of years earlier, Hubbert had predicted in *Technocracy* magazine that attempting to balance the budget would trigger such a crash. Vindicated, in June 1938 he wrote another piece for *Technocracy* about attempts to create forecasts:

As far back as we have record, and doubtless farther, one of the greatest concerns of the human race has been the problem of what is going to happen in the future. So great has been this concern that in most places and times there has existed a separate priestly class whose special function was to make predictions and to give advice, for a consideration, to the officers of state and the leaders of business concerning the probable outcome of any venture that might be embarked upon.

The modern equivalent of prophets, oracles, and seers were "professors and departments of economics," Hubbert quipped, "and in the place of the court astrologer, or the priests at the temple of Delphi, we now have the Brookings Institution." Hubbert felt Technocracy's approach was different. It was scientific.

In his article, Hubbert noted that US population growth had gradually slowed over the past century and appeared to be following an S-shaped curve. So he forecast the current US population of 130 million would grow only slightly, leveling off by the 1950s below 140 million.*

Likewise, Hubbert thought overall economic activity would level off. Echoing what he'd written in the *Study Course*, he declared that for the United States "the period of exponential expansion with present energy resources has already passed." The nation, he thought, had begun to slow down, approaching a stable, steady state.

The *Study Course* had cited estimates for US coal resources and talked about the peaks and declines of resources in general but without trying to forecast the timing. In his latest article, he ventured his first prediction for the nation's oil supplies.

While we do not know exactly how much oil remains undiscovered, we do know that it is a limited supply and that for every barrel of oil taken out of the ground there remains one less barrel to be produced, and that every new pool discovered diminishes the undiscovered pools by one. Furthermore, the easy discoveries have

---

* In a report later that year, titled "Population Problems," the US government's National Resources Committee likewise showed similar S-shaped curves, with population leveling off between 140 and 160 million. "The idea that we are approaching an era of stationary or decreasing population comes as something of a shock to most Americans," the committee stated—but argued this prospect "furnishes no occasion for alarm."

already been made and only the difficult ones remain. While it is improbable that all the oil will ever be taken out of the ground or even discovered, it is certain that the production of oil will reach one or more peaks and finally decline, reaching ultimately the limit of zero production. The time of the beginning of this decline in the United States is somewhat uncertain, yet it seems doubtful that it can be postponed any later than 1950 and possibly not that long.

The article was packed with graphs and analyses—of population, of the quality of iron ore, of the efficiency of coal mining—but Hubbert didn't provide any more detail to back up his expectation of a peak in US oil production by 1950. An in-depth study would have to come later.

KING HUBBERT WAS in a quandary about what he would do after leaving Columbia. Academic labs in America had done little to advance geophysics, he felt—far more progress had come from government and industry labs. He took civil service exams in geophysics and physics, a step toward getting a staff job in the government.

He also considered putting more effort into Technocracy, which was finding success. In late 1937, Howard Scott had made a successful cross-country tour, culminating in a talk at the six-thousand-seat Shrine Auditorium in Los Angeles—after which Hubbert had written proudly to Nell, "We're going to put this country on edge yet baby!" In the fall of 1938, Scott made another cross-country trip, with *Technocracy* magazine reporting that he drew 2,500 in Akron, Ohio, and more than 3,000 in Winnipeg, Canada.

Meanwhile King had gotten divorced from Yelena in 1936, as she had another man to marry, and Miriam had gotten a divorce from her husband the following year. King and Miriam looked into the marriage laws in various states to find out what was required. Many states wanted medical certificates and disclosure of past divorces. Some states would marry nonresidents—but only if the marriage would also be legal in the couple's state of residence. Although they collected information on these statutes, they didn't follow through. But when King filled out paperwork, he began stating that he and Miriam had wed in Atlantic City in November 1938. Even if not legal, they considered themselves husband and wife.

By the spring of 1939, with his time at Columbia coming to a close, Hubbert was still at a loss for a career path. "Am considering getting out of universities for good," he wrote to Nell. "Thinking seriously of taking up a full time job as Director of Research for Technocracy, Inc. That seems to be about the most devastating activity open at the moment. I am a little afraid my country is going to need me if there is any of it left by the time our businessmen and politicians are done with it."

10

Fighting Mad

IT WAS EARLY SEPTEMBER 1939—the same week Hitler invaded Poland, he recalled—that Hubbert came across a notion that to him seemed simply wrong. What began as a discussion of a minor issue eventually turned into a major obsession for him.

It started when Hubbert attended a geology conference in Chicago, which gave him a chance to catch up with friends there, including his former officemate William Krumbein. The two of them got to talking about hydrology—how water moves underground—a subject neither of them knew very well. Krumbein mentioned an equation called Darcy's Law, a relation for calculating how fast fluids would flow through rock, depending on the pressures applied to them.

Hubbert's first impression was that the equation couldn't be right, at least not the way Krumbein stated it. Although meant to describe moving fluids, if you plugged in zero for the velocity of the fluid, the equation gave a ridiculous result: that the pressure was the same everywhere in the fluid. That wasn't realistic. Any fluid on Earth is subject to gravitational pull, which means that when stationary, the pressure at the bottom of the fluid is higher than the pressure at the top—just as water pressure in the deep sea is much higher than it is near the ocean's surface.

In his defense, Krumbein said he got the equation from Morris Muskat's book *The Flow of Homogeneous Fluids Through Porous Media*.

Muskat—a brilliant young physicist working for Gulf Oil—had published this seven-hundred-page book a couple of years earlier, at age thirty. This book had laid out more complex equations, but then in many applications, it had used the simplified version that Krumbein had mentioned. Muskat's assumptions were perfectly reasonable for many situations, such as when fluid was flowing in a horizontal plane, where gravity would be constant.

However Hubbert felt that Muskat hadn't explained his assumptions and reasoning clearly. Hubbert wanted to see a completely general, very rigorous treatment of hydrology—and felt that Muskat hadn't achieved that. Hubbert may have been nitpicking. But he was unemployed, with no work demanding his attention. He decided to develop his own theory of hydrology.

HUBBERT HAD FIRST delved into hydrology a few years earlier, when asked to teach a course at Columbia for mining engineers. He'd become familiar with analyzing electricity from his work on resistivity surveys, and he figured that since electricity is a flow of electrons, then fluid flow could be described mathematically in an analogous way. When a battery pushes a flow of electrons around a circuit, they're said to be driven by a difference in "potential energy" between the positive and negative ends of the battery. So Hubbert developed a way of analyzing the flow of fluids in terms of "potential," as a measure of energy. Then fluids would flow from high potential—such as areas at high elevations or under high pressure— to areas of low potential. Hubbert hadn't seen fluids treated this way in the literature, but he hadn't read much on hydrology.

Years later, when he got talking about fluid flow with Krumbein at the 1939 conference, Hubbert grew skeptical about the rules of thumb used in hydrology. He found that a simplified version of Darcy's Law was widely used in the petroleum industry, having been written into the American Petroleum Institute's code of practices. Although hydrologists often said fluid flowed from areas of high pressure to low pressure, Hubbert could imagine cases where water would flow from one spot to another with the same pressure—or even where it would flow from low pressure to high pressure. According to the usual thinking, that would be like a river flowing uphill.

Hubbert decided to go back to basic physical principles and build a rigorous theory of fluid flow from scratch. Comparing his results against others', he found example after example of what appeared to him as sloppy reasoning and questionable assumptions that had given answers that didn't make sense—sometimes even violating basic physical laws. Sometimes they showed water flowing outward in every direction from a particular spot—as if that spot were generating water from out of nothing. Or they had shown flows that violated the law of conservation of energy—describing, in effect, a kind of perpetual motion machine.

In the midst of this work, he wrote to William Rubey about his progress: "My results are in some important respect or another in disagreement with those developed or employed by almost everyone from the time of the classical studies of Slichter to the present, not excluding such competent theorists as Morris Muskat." Hubbert, unknown among hydrologists, planned to challenge the field's giants.

EARLY IN THE process of writing his hydrology study, Hubbert had contacted Rollin Chamberlin to see if he'd be interested in publishing it in *The Journal of Geology*. Chamberlin had replied he would be. At first Hubbert thought the paper would be about ten pages long. But as he continued working on it, he wrote more and more. When it looked as if his paper would stretch to about fifty pages, he checked back with Chamberlin, who said that length would still be all right.

For two more months, Hubbert wrote version after version of his study—all by hand, since he could only type with two fingers—and it grew ever longer. Miriam, in her time off work as an executive assistant at a medical company, typed up the final version for him. In March 1940 they rushed to get the paper ready for Chamberlin, so he could fill a spot in the journal. As soon as it was ready, Hubbert packaged it up and sent it express to Chamberlin. "I relaxed for the first time in weeks," Hubbert recalled. "My wife and I were settled down for a couple of good scotches and sodas. The wheels started turning around and I found myself doing a little calculation, and finally it suddenly crashed through to me. My God! This thing's going to be over a hundred pages!"

He immediately wrote to Chamberlin. He'd discovered "an embarrassing blunder," he confessed. "I don't know how I made this mistake. I must have done so much writing and rewriting on it that I must be getting goofy." Chamberlin chided Hubbert, saying the stack of paper he'd sent in was as thick as four printed journals. It should have been obvious it was too long.

After an agonizing wait of a few weeks, finally Chamberlain accepted Hubbert's paper. But to print it, Chamberlin would have to arrange for a special supplementary issue of the journal, devoted solely to Hubbert's paper. They gave the study a simple title, "The Theory of Ground-Water Motion"—almost as if it were the first such theory ever published.

As his study went through page proofs, Hubbert checked it meticulously. He expected, he told Chamberlin, that "a sizable corps of interested critics are going to comb this pretty carefully for errors."

WHILE WAITING FOR his groundwater paper to appear in print, Hubbert contemplated what to do next. "My experience at Columbia has been a kind of hell and I am washing my hands of it shortly," Hubbert confided to Carl Eckart, a physics professor at Chicago whom he liked and respected. "Personally I have done so much thinking during the past 10 years," he added, "that I think I would be judged by any 'sane' person to be definitely 'insane.' And, I might remark that Columbia University has the highest percentage of 'sane' people I hope I shall ever have to deal with."

Writing to Eckart, Hubbert was unusually forthright and caustic. "I am still interested in teaching if I can find a set-up that will allow me to do so without being continuously frustrated by a bunch of colleagues and superiors who are so dumb they hardly know which way is up. Without such a set-up I am definitely through with academic work. Life is far too short to be wasted on such futility."

Hubbert wasn't ready to give up on science, though. Over the past dozen years, he'd compiled a huge mass of information on geophysics for his classes, and with his papers on scale models and groundwater, he felt he'd made significant contributions. He wanted to take all this material, he told Eckart, and "synthesize it into something of a treatise." A month later he also wrote to Rollin Chamberlin about his plans, saying he would "hibernate" for a year and churn out a geophysics book. That is, he added, assuming "all apple carts are not upset by this international mess."

The war in Europe had recently become decidedly grimmer. In May 1940 a German blitzkrieg forced the surrender of Belgium, then in June, of Norway and France. That left Britain the only country in western Europe still in the fight against the Nazis and Fascists. Roosevelt wanted to enter the war, but many Americans hoped to avoid another bloody debacle like the Great War.

At the time, Hubbert and the Technocrats opposed US involvement. "I didn't feel it was our war," Hubbert said a couple of years later. "I was hopeful we might be able to keep out of it."

HUBBERT'S BOOK-LENGTH "Theory of Ground-Water Motion" finally appeared as a special issue of *The Journal of Geology* in December 1940. A couple of months later Hubbert gave a talk on it at the annual conference of the American Institute of Mining and Metallurgical Engineers. Morris Muskat was there, as well as researchers from the Geological Survey. Instead of trying to collaborate with established hydrologists to improve the theory of groundwater, Hubbert launched an attack on them, harshly criticizing their work.

Following Hubbert's talk, the survey's researchers were incensed, and Muskat also challenged him, getting "emotional and abusive," as Hubbert recalled. He managed to make them all "fighting mad." Neither side pointed to any significant mistakes by Hubbert, but they felt he was nitpicking and that the theories he criticized were actually correct within the domains they had addressed.

The way scientific ideas evolve, Hubbert once put it, is "if you are wrong, you are buried—and if you are not wrong, you will probably survive." It wasn't clear whether Hubbert's ideas would survive.

IN THE EARLY months of 1941, the war in Europe continued to worsen. By May, Roosevelt declared an "unlimited national emergency." In July, the United States placed an embargo against Japan, blocking sales of steel and oil—crucial, since Japan had no oil resources of its own. As Hubbert had written in a Technocracy memo several years earlier, studies of strategic materials could answer questions such as "Where does Japan obtain the oil she requires to operate her battle fleet? What would happen if she were blockaded?"

On December 7, 1941, Japan answered those questions, launching a devastating surprise attack on the US naval base at Hawaii's Pearl Harbor. The following day the United States declared war on Japan, and in turn, Nazi Germany declared war on the United States.

Although Hubbert had hoped America could stay out of the war, "it didn't pan out that way," he reflected several months after Pearl Harbor. "In other words, I was wrong."

Total Mobilization

BY THE TIME OF THE Pearl Harbor attack, Hubbert had been without a job for two years. Ever since his fight with Columbia, he'd been looking for other work. Although he'd passed the civil service exams for both physics and geophysics, he hadn't secured a government job. He also had a connection with Bell Laboratories in New Jersey, where a friend of his worked on computing machines, but nothing panned out there. Frustrated, Hubbert wrote to William Rubey that his job search had resulted in little except "many false alarms."

With the start of the war, however, Hubbert's job prospects seemed to improve. The war mobilization was boosting the economy. Factories were humming again. The government ballooned, adding new administrations and boards and commissions. Like tens of thousands of others, Hubbert ventured to Washington, DC, in search of a job. He had a long list of contacts to try. Some he knew well, like the economist Leon Henderson. Others he'd spoken to briefly, and a few he didn't know at all.

He arrived in January 1942, which happened to be a bad time for a Technocrat to seek work in the federal government. A loudmouthed but intimidating congressman named Martin Dies was just then attacking Henderson for his past ties with Technocracy.

Martin Dies represented Texas's 2nd District outside Houston, home of the Spindletop oil gusher discovered back in 1901. As a Democrat,

Dies had initially been a strong supporter of Roosevelt's New Deal. But by Roosevelt's second term, like many southerners, Dies had soured on the president. By the time the United States entered the Second World War, Dies had become the nation's leading crusader against Nazis and Fascists, Communists and their "fellow travelers"—any kind of "subversive elements." Dies chaired the House Un-American Activities Committee, often called simply the Dies Committee, a position he used to target officials who were responsible for New Deal projects or otherwise close to the president.

By late 1941, Leon Henderson had become "the New Deal's number 1 economist," as one magazine put it, and was one of Roosevelt's most devoted aides. He was in charge of the Office of Price Administration, part of the Department of the Treasury, which had the thankless job of setting prices of a multitude of goods—both to fight inflation spurred by sudden wartime economic growth, and to ensure civilians didn't use up materials the military needed. Henderson became one of the most public faces of the burgeoning war effort and got dubbed America's "price czar." With his huge paunch, and his habit of constantly chomping on cigars, he looked the part.

A few months before the United States entered the war, Dies went after Henderson, writing to the president, "Leon Henderson has surrounded himself with highly paid assistants who are, by their own public records, strangers to the American way"—a code phrase for subversives. Henderson shrugged it off, vowing to "eat on the Treasury steps any Communist organization to which I belong."

In January 1942, Dies launched a new attack, charging that Henderson had been a member of Technocracy, "one of the craziest economic propositions that was ever hatched in a crackpot's brain." Dies also pointed out that one of Henderson's analysts was Harold Loeb, a writer who had published the 1933 book *Life in a Technocracy*, inspired by discussions with Howard Scott.

As a profile in the magazine *Collier's* had put it, Henderson was "the loudest, fightingest, busiest man in Washington." Here he lived up to that reputation. In response to the new charges, Henderson said he'd often enjoyed discussing issues with Scott but had never been a member of his group. "My offer still stands good," Henderson said. "I will eat on the Treasury steps any subversive group I belonged to. Try again, Mr. Dies."

Dies turned to other targets. But there was no denying Loeb's Technocratic ties. Henderson let Loeb go.

WHILE DIES PUT Technocracy in the headlines for the first time in years, the group launched a new campaign that would also make news. A month earlier, immediately after the Pearl Harbor attack, Howard Scott had sent a telegram to President Roosevelt, pledging the assistance of all Technocrats. Then the organization launched a letter-writing campaign calling for Scott to be named "Director-General of Defense," to oversee the nation's war efforts. Over the next few months, the White House received thousands of such letters from Technocracy sections nationwide. Meanwhile the organization also placed a series of full-page advertisements in newspapers across the country, likewise nominating Scott to lead the war effort.

The ads triggered an immediate backlash. In editorials and letters, Technocracy was called Fascist and totalitarian. Following these attacks, Technocracy's high-profile ads continued, appearing in *The New York Times* and *The Washington Post*—but with a crucial change. There was no call for a director-general of defense, no mention of Scott at all. They simply urged "Total Mobilization."

When a reporter asked about the earlier ads, Scott blamed them on over-enthusiastic Technocrats in various local offices. "This is Continental Headquarters," Scott said. "We know nothing of what the regional governors do." Scott didn't mention the memo from Continental Headquarters that had kicked off the campaign.

HUBBERT REACHED OUT to Leon Henderson's office but didn't get a job offer. He tried a contact at the National Resources Planning Board, the nation's main advocate for centralized planning. At the Office of

Production Management, he tried Morris Cooke—who, the year before, had helped him get an article on Technocratic ideas published in the journal *Advanced Management*. Writing to Rubey, Hubbert told of how he'd tried "crashing the gate" at the Office of Scientific Research and Development, hoping to talk to its director, the engineer Vannevar Bush, who also served as the president's science adviser.

For weeks, none of Hubbert's leads came to fruition. Finally he got some traction with the Bureau of the Census, which wanted to hire him to track resources but lacked the budget. Census put him in touch with the Board of Economic Warfare (BEW), a fairly new agency run by Vice President Henry Wallace. "Economic warfare" could involve designing embargoes, like the one against Japan, as well as identifying resources around the world and buying them—either because the Allies needed them, or to keep them out of enemy hands. Such work on resources was what Hubbert had been looking for.

In April, the BEW offered him a position, and he accepted. He'd be a senior analyst on world mineral resources, making $4,600 a year—more than double what he'd earned at Columbia.

A FEW WEEKS before Hubbert joined the BEW, Martin Dies had turned his sights on that agency, charging that "at least 35 high officials" employed there were subversives. He named several but directed special attention to the BEW's principal economist, Maurice Parmelee.

A gray-haired, bespectacled fifty-nine-year-old, Parmelee appeared harmless enough. In a wide-ranging career as a professor and author, he'd written the first modern textbook on criminology, as well as books on sociology, politics, and economics. In his 1935 book *Farewell to Poverty*, he had argued, "The high technological development in the United States renders it feasible to introduce a planned social economy much more readily than has been the case in the USSR." America, he seemed to be saying, was more fertile ground for socialism or communism than Russia.

In attacking Parmelee, Dies brought up these notions, but he focused more heavily on another part of Parmelee's thinking: nudism. Parmelee was no secret nudist. He'd written a book advocating it, *Nudism in Modern Life*. Published in both Europe and the United States, it had become the subject of a censorship trial.

Dies's revelation spurred Congress's first-ever debate on "life sans clothing," *The Washington Post* reported. The Library of Congress, it turned out, had both the first and second editions of Parmelee's nudism book, which a Georgia Democrat carried into the House, declaring, "If I have ever seen in print and picture anything that is filthy and dirty, they are these books." His fellow congressmen eagerly passed the books around. A favorite image depicted a young woman throwing a spear, wearing nothing but a necklace.

After Dies's attack, the BEW let Parmelee go, using the pretense of "reorganization."

AT THE BEW, Hubbert was charged with building a small group of researchers to compile information on resources across the British Empire. But it was tough to find good geologists, since most of them were either off fighting or already working for another agency. With the help of friends around DC, including Rubey at the Geological Survey, Hubbert managed to assemble a solid team.

The Civil Service Commission, responsible for background checks on new government employees, was running behind. A few months after Hubbert started at the BEW, two investigators came to talk to Hubbert about his past, in particular his involvement with Technocracy. He cooperated fully, telling them about the organization and giving them stacks of publications. The investigators, being very thorough, also visited Technocracy's headquarters in New York.

Hubbert never heard back from Civil Service and went on with his work. Most of his team's assignments came directly from the army, and they churned out reports with titles like "Strategic Minerals from India, Ceylon and Afghanistan Which Could Be Transported by Air."

After Hubbert had been at the BEW more than four months, Civil Service called him in for more questioning. When asked about Technocracy, he was forthright. The group's view, he said, was that "science and engineering are able to provide a better standard of living and to operate the country better than the conventional methods by which it is done now." Figuring out how to supply the nation's needs involved solving "technical problems," as Hubbert put it, and he asserted that Technocracy's approach was "entirely within the framework of the present government and present constitution."

However, he was critical of the country's current governmental setup. It lacked "the technical brains to run the country," he told the investigators. "There is no device in our political system to get men of that kind" into positions of power.

HUBBERT PUT IN long hours, working six days a week, but occasionally he found time for breaks. Once, on a Saturday, he took one of his assistants to a Negro League baseball game, which whites rarely attended, and they got to see the famous pitcher Satchel Paige. The nation's capital, like most of the South, was deeply segregated, so blacks were barred from most jobs, even from using most businesses. It was a far different world than Hubbert's favored Manhattan neighborhoods, where all races intermingled.

In early 1943 the investigation into Hubbert's past was still unresolved. "There has not been another squawk out of the Civil Service Commission, and I do not expect anything further," he wrote Howard Scott in early 1943. "The only thing that could cause trouble now that I can see would be for the Dies Committee to sound off, but the latter has been rather subdued of late."

Just weeks after Hubbert wrote that letter, Dies did sound off, presenting a list of forty alleged subversives in government. Dies shouted for Congress to withhold agencies' funds until they purged any "irresponsible, unrepresentative, crackpot and radical bureaucrats"—and got a standing ovation from both sides of the aisle.

In this new round of attacks, Dies went after Harold Loeb again.

After being forced out of the Office of Price Administration, Loeb had popped up at the War Production Board. This time Dies delved deeper into Loeb's past. In addition to *Life in a Technocracy*, Loeb had written a 1936 book, *Production for Use*. It had concluded that the nation faced two possible paths: capitalism could be saved through massive war spending, or it would collapse and be replaced by some kind of technocratic state. Whichever path America followed, Loeb argued, "we may as well dismiss any idea of returning to the open market system."

"I know of no man," Dies raged, "who could better qualify for the title of perfect crackpot."

SEVERAL WEEKS AFTER Dies's latest attack on Technocracy, the Civil Service Commission finally made a decision on Hubbert's case, writing to the BEW that he had been "rated ineligible" because of his involvement with Technocracy. The BEW was "requested to separate him from his position." Rather than give in, Hubbert's bosses at the BEW decided to hold their own formal hearing.

When Hubbert came to BEW headquarters for his hearing on the morning of April 14, 1943, he reported to Room 2240. There to question him was Thomas P. Brockway, a special assistant in the BEW. Brockway had been a Rhodes scholar and earned his doctorate in history at Yale, and before joining the BEW, he had written a book, *Battles Without Bullets*, explaining economic warfare to the layman. (Incidentally, Brockway's middle name was Parmelee; his uncle was the nudist scholar Maurice Parmelee.)

"I have no particular plan for starting," Brockway said disarmingly, "but I thought we might begin by asking Mr. Hubbert to tell us what Technocracy, Incorporated, is, and what its aims are."

Hubbert had his own ideas of what to focus on. The charge, he said, was that he'd been a member of Technocracy, a group that had the same legal standing as any number of scientific organizations that he was also a member of. "Put in that form," Hubbert said, "it is no charge at all."

It was true that, legally, Technocracy could say almost anything it liked. However a 1939 law, the Hatch Act, had barred the government from employing members of political organizations that advocated overthrowing the government. (On starting at the BEW, Hubbert had filled out a form attesting that he was not a member of any such group.) The attorney general was actively compiling a list of organizations deemed subversive or suspected of being so—from the notorious Communist Party of the USA down to the Central Patriotic Committee, a one-man outfit in Witchita, Kansas, that distributed pro-Fascist literature. Though Technocracy wasn't on the list, Civil Service thought the organization highly suspicious.

To try to defuse the charges, Hubbert explained Technocracy's overall goals. "The primary interest of the organization is simply a high standard of living, high standard of public health, minimum wastage of nonreplaceable materials," he said. "That is just about the works."

"What is your own view of the best way to achieve the aims you have stated?" Brockway asked.

"I would say predominantly education," Hubbert replied. "We are undergoing an entirely spontaneous and entirely automatic evolution which nobody can stop." The best way forward was to have intelligent people debate, and "by a give-and-take process and an interchange of ideas, something finally emerges as the better solution." That was his "own personal notion of how social evolution takes place."

"How do you visualize the situation in this country politically, economically, and socially?" Brockway asked. "Do you expect to find pretty drastic changes in our institutional setup?"

"Whether I exist or whether Technocracy exists, either one," Hubbert replied, "I expect quite marked changes in the institutional setup in the next twenty years, quite as marked as have happened in the last ten years and possibly more so."

"Do you expect the electoral system to be radically changed?"

"It has been in the past and I expect it will be in the future," Hubbert said. He asserted that the government had grown into a behemoth bureaucracy, which the public had never directly voted for. "We carry out suffrage as a sort of fiction. It is a device for appearing to have popular election."

"Do you think that some other actual power should be brought into Government than the present power?" Brockway asked.

"No," Hubbert replied. "The only question that is involved is an optimum method of organization. If you want to know if I think the existing Federal Government is the optimum government organization, the answer is 'no.'"

"Would you like to indicate what would be your idea of the best arrangement?" Brockway asked.

"Why, the best arrangement," Hubbert said, would be to run industries efficiently, to "keep that equipment running and to deliver its product to the people without getting stopped periodically and without one part of it getting balled up and gumming up the other part." It would be coordinated so that when managers "decided to do a certain thing they could at least get it done."

"How would you get your engineers into the position that they could do that job?"

"When the country gets into a bad enough situation, when it wants them."

"When it wants them, how does it express that?"

"When the people of the United States decide they want that sort of thing, I have no doubt they will take practical action for getting it."

"Through the ballot?"

"Presumably."

"You mean they will vote for engineers?"

"They might vote for a constitutional amendment."

"What would be the nature of the constitutional amendment?"

"Simply the adoption of a new Constitution."

"The government you would like to see would be a government with more power than the present one?"

"Decidedly so," Hubbert replied. It would "combine in the United

States Government powers held by the big corporations. I sometimes wonder who is more important, Standard Oil or the United States Government, for a perfectly good reason. A big corporation is a very effective organization, but it is not operating in the interest ordinarily of the citizenry. In other words, I would like to see a government that truly represents the public on one side and on the other side had the power to really get things done."

Here, Philip Dunaway, a BEW assistant sitting in on the hearing, spoke up: "What additional power could you give the United States Government that it now lacks?"

"Give it the power to run all of the industrial equipment that is now run by so-called private industry," Hubbert replied.

"How would you run the oil business in this country under the plan you are describing?" Brockway asked.

"Simply have the oil division of the government in charge of it," Hubbert said. This oil division would figure out how to supply oil and also tackle policy issues, he said: "What are we going to do with the oil? Shall we burn it up at this rate? Should we limit it only to these special uses, use coal for something else?"

Hubbert's questioners backed up to get a broader picture. "Does what you have said here today reflect your views?"

"It reflects some thinking that I have done, yes," Hubbert said. "And as far as I know, thinking is not illegal."

The BEW investigators asked about Hubbert's work for Technocracy, and he said he'd written "a few odds and ends for the literature of the organization" (but didn't mention that he'd written their main text, the *Study Course*).

Dunaway asked about one of Hubbert's articles, "Economic Transition and Its Human Consequences," in the journal *Advanced Management*—the article that Cooke at the Office of Production Management had helped him get published a couple of years earlier. This article laid out many of Technocracy's ideas, arguing the nation had so far staved off collapse of the price system through a variety of means, such as the advent in the 1920s of widespread consumer debt, and the "pyramiding of the federal debt" under the New Deal. The article concluded that government deficit spending was "only a temporary expedient, yet it is the only thing at present that is keeping the country off the rocks." (Since he'd written this, wartime spending had also given America's economy a huge boost.)

"Could a man who read this article form a judgment as to what your views were?" Dunaway asked.

"I certainly did not write that article with my fingers crossed," Hubbert said. "When I sign a piece of paper, I mean it. That goes for the record. I do not write things I do not believe."

Hubbert's boss, an economist named Joseph Gould, was also attending the hearing. Brockway asked Gould about Hubbert's work. At the time Hubbert was studying "the types of minerals and fuels that would be needed in Greece and in Italy in the event of occupation," Gould explained. "Both are very important studies." The Army Corps of Engineers had sent letters of commendation for Hubbert's reports on fuel resources and building materials, he added. "They said they were so good that instead of reworking it themselves they have incorporated it in their handbooks."

"Would you care to say a few words on the effect of your operations in case you lose Mr. Hubbert's services?" Brockway asked.

"If we lose Mr. Hubbert's services," Gould said, "our present Minerals and Fuels Unit is practically shot to pieces."

"Would you have difficulty in replacing Mr. Hubbert?" Brockway inquired.

"Yes," Gould said, arguing it was difficult to find people with the right expertise, who understood how to pinpoint mineral resources around the world and, ultimately, what was physically possible to achieve. "You cannot get all economists to do that because of their training," he explained. "You do have to have somebody with a scientific background."

Brockway asked whether Hubbert had any concluding statements. He said he understood the BEW's position and that investigating the charges against him was "the only honest and intelligent thing to do." But he was derisive about the Civil Service's "dumb investigators." He added,

"These boys are just trying to make a job for themselves. That is the only sense we can make out of it."

HUBBERT HAD TAKEN an adversarial approach during the questioning. But talking with Brockway off the record soon afterward, he realized that actually the BEW had been trying to clear him. The hearing had been held not to condemn him but to give him a chance to show that Technocracy was not a threat. He wrote Brockway an apologetic letter, elaborating on Technocracy's ideas and aims. The BEW continued fighting to keep Hubbert, writing to Civil Service to request an explanation of exactly what was objectionable about Technocracy and why his involvement in the organization was considered a problem.

In response, Lawson Moyer, the Civil Service Commission's executive director and chief examiner, sent a long letter with excerpts from a report on Technocracy from the Office of Naval Intelligence, one of the nation's main spy agencies.

The excerpt from the intelligence report began, "The dominant figure in Technocracy, Incorporated, is one Howard Scott, and Mr. Hubbert is said to be the power behind him." The report listed various scraps of information about Technocracy, most of which had been reported in newspapers over the previous decade. The report drew parallels between Technocracy and the Nazis. "Is it true that the Technocrats have no plan of action?" the report asked. "If it were true then the situation would be as if Hitler wrote 'Mein Kampf' as a blueprint and stopped there." The report argued that Technocracy was "mobilizing their own 'Storm Troopers'"—apparently a reference to its members' penchant for gray suits and its caravans of gray cars that accompanied Scott on his speaking tours. "This may sound bizarre," the report admitted, "since scientists and intellectuals are not given as a rule to such activities." The report also called the Technocrats' monad symbol "reminiscent of Hitler's swastika."

In his letter attached to the intelligence report, Moyer stated, "The above is self-explanatory and it is believed that it will furnish the Board of Economic Warfare with all the information requested."

. . .

MEANWHILE, THE BEW became the center of a public battle between its head, Vice President Henry Wallace, and another top official, a Texan businessman named Jesse Jones. The BEW had never fit easily into the Washington bureaucracy, as its role overlapped with those of well-established agencies and departments, in particular the Reconstruction Finance Administration, which Jones had run for nearly a decade.

Jones had made a fortune in banking and real estate and seemed to own half of downtown Houston. Early in the Depression, he had come to Washington to help Hoover's administration bail out failing banks and corporations. When it came to fighting the Second World War, he thought it essential to strike good deals, to avoid getting ripped off, and above all to keep from going broke. He had come to play such a central role in the nation's financial decision-making that, as *The Saturday Evening Post* described him, "next to the President, no man in Government and probably in the United States wields greater power."

Hubbert had heard a story about Jones—one that went around the capital—that perfectly captured his money-minded outlook. Back in 1941, when the United States imposed its embargo against Japan, Japan had retaliated by cutting off US access to Southeast Asia, home of the world's main rubber plantations. No one had developed a high-quality synthetic form of rubber, so the United States scrambled to stockpile as much rubber as it could—an effort Jones oversaw. One night a fire broke out that burned up sixteen thousand tons of stockpiled rubber. The material was priceless—in the sense that no matter how much money the United States had, it couldn't simply go out and buy more. Yet when told of the bad news, the story went, Jones brushed it off, saying, "Well, it was insured, wasn't it?"

Even after the BEW appeared on the scene, Jones still held the purse strings for imports of materials to build up the war machine, so Wallace and Jones had to work together. But Wallace took a starkly different view of the war effort. Wallace was an Iowa farmer and scientist who had founded a successful corn corporation. He'd then served as the secretary of agriculture and come to prominence as the architect of the New Deal's farm relief programs. He also had a mystical side, seeing the Second World War in stark terms, a battle of good against evil over the world's soul. Wallace took the view that the nation must get the materials needed to win the war, regardless of the monetary costs.

Throughout the war, Wallace felt Jones was a penny-pincher who'd hampered America's economic warfare, sometimes allowing their enemies to outbid them in international auctions for resources. Jones rejected this criticism, arguing the nation had been stockpiling plenty. After fighting a turf war for two years, Wallace finally snapped. In the summer of 1943, he gave a major speech accusing Jones of "obstructing the war effort" and of following "timid, business-as-usual procedure" and released a twenty-eight-page report detailing his grievances. Jones shot back with a thirty-page rebuttal. President Roosevelt had warned his officials not to fight in public, to avoid undermining public morale. When this internal battle wound up in the newspapers, Roosevelt quickly removed Wallace as director of the BEW and also stripped Jones of much of his power. It was a major demotion for both combatants.

When the news broke in mid-July, Hubbert wrote to Howard Scott, "They seem to have blown a couple of lids off!" By that time, he hadn't heard much more out of the Civil Service Commission. He thought the investigation had fizzled.

But Dies popped up again, going after another of Hubbert's colleagues at the BEW—Parmelee's replacement. This man, John Bovingdon, was also a dancer who had advocated "a new school of rhythmic expression through the ballet," as the *Chicago Tribune* put it—and had allegedly danced in support of Soviet Russia.

Within days of the Bovingdon affair, Civil Service renewed the pressure to fire Hubbert. By then Hubbert was losing his will to fight. He was "disgusted" by the fighting among agencies, especially the battle between Wallace and Jones. He wanted out of the government.

PART 3

Houston,
1943–1956

No Doubt

BY AUGUST 1943, HUBBERT FELT he was close to breaking into the oil industry, which would give him a way to escape Washington. He'd interviewed with Standard Oil at its headquarters in New York City, and while there he'd also met with another major oil producer, the Texas Company. Then his friend Darrell Hughes at Shell Oil, the US branch of the European oil giant Royal Dutch Shell, sent a cable saying he had "a very interesting proposition" for Hubbert.

Almost a decade earlier, Hubbert had given Hughes a job on his electrical resistivity crew, then had put him up in his apartment until he got a permanent job. Hughes had gotten a start in the oil industry and eventually worked his way up to a position as head of Shell Oil's geophysical lab, based in a suburb west of Houston.

In a position to repay past favors, Hughes arrived in Washington in early September with welcome news. Shell had a job offer for Hubbert, an unusual position as a technical coordinator. Hubbert's task would be to keep abreast of the latest studies appearing in journals, as well as what Shell's researchers were up to, and then inform the company's researchers about new findings relevant to their work. He'd be a sort of walking library or scientific matchmaker. He'd get to "keep in touch with a very broad range of things," he later recalled, "rather than have my nose stuck in some little two-bit problem or other."

The main downside was that the Shell job was in Houston. Although Hubbert had grown up in Texas, he'd rarely returned to visit his family, and he certainly didn't want to live there again. But Standard Oil didn't offer him anything in New York. He'd ruled out more government work and had given up on academia. So he quit the BEW and somewhat reluctantly took the job at Shell.

Although Hubbert didn't know it, Shell's offer had come just in time. The company had a policy of only hiring researchers under forty years old, and Hubbert's fortieth birthday was just three weeks away.

STARTING AT SHELL, Hubbert felt like a nobody, as if he were there only on a trial basis. Rather than landing the type of job Hughes had described to him, Hubbert was assigned another role. "There was a bit of confusion apparently about that," he recalled. "Different people had different ideas of what I would do." Instead of the technical coordinator position, he was tasked with integrating results from three disparate geophysical techniques. Seismographs, like those he'd used in work for Geophysical Research Corporation, could provide a fine-grained view. Gravity measurements, like those he'd analyzed with Frank Melton, could reveal larger features underground. The industry also used magnetic measurements, which could be done from an airplane flying over a landscape, providing a broad picture of the hard "basement" rock that underlay any softer rock layers at the surface. Hubbert's task was to figure out how to take readings from these very different types of measurements and compile a picture of the invisible landscape underground.

Despite the initial confusion about Hubbert's role at Shell, he liked his managers' style. "Fortunately they had sense enough to leave me alone," he recalled. In Shell's downtown skyscraper, the company gave him an office on the twenty-first floor, immediately below the executive suites. He holed up there so he could "work like hell." He relished the opportunity. He didn't have to teach, didn't have to manage anybody, and didn't have to worry about his next paycheck.

As the oil capital of the world, Houston's economy had remained buoyant through the Great Depression. America's thirst for oil had continued increasing most years in the 1930s, rising from 2.5 million barrels a day just before the 1929 crash to over 4 million barrels a day in 1940. As *Fortune* magazine put it, Houston was "the city the Depression missed." Then, with the onset of war, oil consumption had surged even higher, and the United States supplied six of every seven barrels the Allied forces used.

When Hubbert first moved to Houston, the coastline at night was lit up by hundreds of flares—excess natural gas being burned at the wellhead, a surplus not profitable to capture. Huge petrochemical plants around the city churned out lubricants and other petroleum products. Its shipyard and steel plants were also booming. Workers poured into Houston, but with the war effort consuming so many resources, few new houses were being built, and there was a housing crunch. "The only place that we could find to live was a little auxiliary building that I think had been a chicken house," Hubbert recalled. In Houston's almost tropical climate, the poorly constructed house suffered from a buildup of condensation inside. Through his early months at Shell that winter, Hubbert suffered chronic nose and throat infections. He considered it "a hell of an environment."

On starting at Shell, Hubbert had quickly become consumed with work, but he also found time to get involved in local causes. It was common to donate to the local Community Chest, an organization that redistributed funds to various charities. He wanted to help the Maternal Health Center, a clinic that provided birth control to the city's poor—both white and black, despite the city's rigid segregation. But the Maternal Health Center wasn't on the Community Chest's receiving list. The city's Catholics disliked that the center provided birth control, and the city's white donors may have avoided it because it helped minorities. Whatever the reason, Hubbert felt it was being deliberately excluded. He informed the Community Chest that, rather than give to them, he'd donate directly to the Maternal Health Center.

.   .   .

IN EARLY 1944, with Hubbert still engrossed in his work of integrating geophysical measurements, his colleague Jim Bugbee asked him to give a talk at the monthly dinner banquet of the local chapter of the American Institute of Mining and Metallurgical Engineers. Hubbert saw the talk as his "debut," an opportunity to make a bold impression with his colleagues at Shell and throughout the industry.

First, Hubbert had to get approval from his superiors, and Shell had a reputation for being tight-lipped. Bugbee suggested he speak about his work on scale models. Soon after that paper appeared in print in 1937, Gulf Oil had sent a man to New York to pick Hubbert's brain—so one oil company, at least, had found the ideas intriguing. For his talk at the banquet, Hubbert proposed a more circumscribed topic: the strength of rocks. That subject "had no smell of oil, and no commercial implications," Hubbert figured. Shell was "in a weak position to say no, so they said yes."

The audience would be almost entirely petroleum engineers, so Hubbert figured it best to focus on practical matters rather than esoteric theory. He planned the talk carefully, providing specific examples of how rocks could bend and flow, including a thought experiment about the land that supported them: the state of Texas. Although much of the state was covered with soft oil-bearing sediments, the main source of the state's wealth, underneath those sediments was bedrock. Suppose, Hubbert said, you cut out a block of this bedrock, following the state's borders, to create a huge slab several hundred miles across. And suppose you had a gargantuan crane. Could you lift the state of Texas? "Will the rock itself," Hubbert asked, "be strong enough to permit hoisting in this manner?"

To figure out whether it would be possible, you could scale this situation down to a tabletop model. But to ensure it was accurate, as Hubbert had shown in his 1937 paper, the strength of the materials would also have to be reduced—so the scaled-down "rock" would have the consistency of toothpaste. He asked the audience to imagine sticking some bolts into a layer of toothpaste, rigging up some cables, and then pulling on the cables. "The eyebolts would pull out," Hubbert explained. "If we should support it on a pair of sawhorses, its middle would collapse. In fact, to lift it at all would require the use of a scoop shovel." Poking fun at his

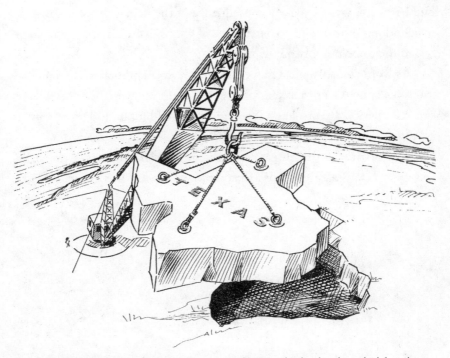

The strength of rock: Hubbert imagined cutting out the bedrock underlying the state of Texas and trying to lift it with an enormous crane.

home state's self-image, he quipped, "The inescapable conclusion is that the good state of Texas is utterly incapable of self-support!"

At the end of Hubbert's talk, Paul Weaver, a polymath who served as Gulf Oil's chief geologist, stood up to comment. He praised Hubbert's work on scaling, explaining the company had found it of great use. For one thing, they'd used Hubbert's concepts to create more realistic models of salt domes. (For decades, buried salt domes had been important sites to drill for oil, and the industry had been developing better techniques for finding them and for modeling how they had formed.) Although Hubbert had tried to pick a topic without applications to oil, Weaver outed him.

Aside from Gulf Oil, though, many in the industry seemed unaware of these laws of scaling. After his talk, Hubbert received many requests for advice. One caller wanted to know: When a long drill pipe is pulled out of the ground, why does it flop around like telephone wire? That,

Hubbert said, was likewise a consequence of the principles of scaling. To make a long pipe as rigid as a short one, the walls of the pipe had to be scaled up, to be proportionally thicker.

As word spread about Hubbert's talk, the local branch of the Geological Society of America asked for a repeat performance. Then the American Association of Petroleum Geologists invited Hubbert to go on tour as a distinguished lecturer, delivering his talk in several states.

All this interest made an impression at Shell. Hubbert's probation was over. "I was all right from that time on," he recalled. "There was no doubt."

13

The Million-Dollar Lab

EVERY OIL COMPANY WANTED TO find more oil to replace its fields, which were continuously depleting, and to boost production as fast as possible to gain a larger share of the growing market. As the US branch of Royal Dutch Shell, Hubbert's employer had the task of trying to find more oil in North America—which was already poked full of far more exploratory holes than anyplace else on Earth. When new opportunities opened up in Texas and along the Gulf Coast, Shell had lost out on lucrative leases, falling behind its competitors.

To catch up, Shell needed to innovate. To enhance its research on exploration and production—E&P, in industry parlance—it decided to greatly expand its small geophysics lab, where Darrell Hughes and a dozen or so other researchers worked. In the Houston suburb of Bellaire, Shell would add a world-class center for some 150 researchers, at a cost of a million dollars—a price tag that made a headline in *The New York Times*.

In the spring of 1945, Hubbert interviewed for a position in the new lab. A couple of months later, he got word that he'd been promoted to become the new lab's associate director. Its head would be Harold Gershinowitz, a chemist working on the refining side of the industry. Hubbert would be in charge of research on exploration and production, making him the lab's top geologist. It was a huge break for Hubbert. In less than two years, he'd gone from embattled government employee to one of the top researchers in the entire oil industry. "The only drawback," he wrote to Howard Scott, was that he was "still going to be located in Houston."

Hubbert and Gershinowitz had the task of designing the lab from scratch—setting its scope, hiring its staff, even helping lay out the floor plan of the 40,000-square-foot building. Hubbert began recruiting researchers in the fall of 1945. Italian leader Mussolini had recently been captured and killed, and Hitler had committed suicide. Soon after the United States dropped atomic bombs on Hiroshima and Nagasaki, Japan surrendered. Once the grueling six-year war was finally over, Shell wanted to recruit researchers quickly. A couple of days after Japan's surrender, Hubbert wrote to his friend William Rubey at the Geological Survey for help finding men to hire, "since almost all of the otherwise footloose geologists in the country are either in the armed forces or else in civilian war work."

Although the new lab would ultimately have to pay off by helping the company find and extract more petroleum, Shell took a long-term approach by supporting fundamental research. "No expense will be spared in the provision of tools and laboratory facilities," Hubbert told Rubey. They'd maintain demanding standards. "In addition to the desired formal training," Hubbert said, he wanted men with "unquestionable scientific and research ability." He reached out to a geologist at Columbia, explaining that the Shell lab would also hire physicists, since "there will be no water-tight compartments within the organization." Above all, he wanted researchers who could think, both critically and creatively.

HUBBERT HAD BEEN worried for years about how the United States would find enough oil to keep its production rising. Even before joining Shell, he'd begun developing a notion about a new kind of reservoir for oil, one nobody had thought to look for. At the new lab, he was finally in a position to put his idea into action.

The earliest oil explorers, with little understanding of where oil might collect underground, had targeted their drilling by looking for obvious clues on the surface, such as seeps of oil or tar. By the late 1800s, the industry had gained more understanding and developed a theory that oil collected in anticlines—arches of rock that trapped buoyant oil underneath. Later, geologists discovered that oil could also collect in more complex types of traps, on the flanks of salt domes and along the edges of faults.

Hubbert argued for the existence of another type of trap—a type no one had explicitly searched for yet. Oil had a tendency to migrate upward through rock, but it could also be pushed around by groundwater flows—and Hubbert thought these opposing forces could sometimes strike an unusual balance. When one is driving a car through an intense rainstorm, water can pool on the windshield, balanced between gravity and wind. Similarly, where groundwater flowed through rock, it could make oil collect in unexpected places, Hubbert figured, a phenomenon he called "hydrodynamic entrapment."

This idea had grown out of Hubbert's 1940 study on groundwater, and he had developed the basic ideas while out of work, before joining the BEW. On starting at Shell, he'd wanted to bring up his idea and encourage the company to put it to use—but he felt it wouldn't get much traction. Once he became assistant director of the Bellaire lab, he began feeling things out, learning that there was a long-running rivalry between two sides of Shell's research: the exploration folks and the production folks. Hubbert came from the exploration side, while hydrology was usually handled by the production side. If he brought up his idea for hydrodynamic traps, he figured he would be "on somebody else's turf, and that would be intolerable."

He finally found an opening in the summer of 1946 at Shell's annual exploration meeting. Most of the attendees were vice presidents and exploration managers, responsible for deciding how to spend tens of millions of dollars on drilling and refining. Hubbert and Gershinowitz also came, representing the new E&P lab. During the meeting, Shell's exploration manager for the midcontinent region, Sherwood Buck-staff, explained a problem they'd encountered. In drilling on the flanks of Oklahoma's Arbuckle Mountains, wells were coming up "dry"—meaning they found no oil, although as with most deep wells, they did hit water. Exploratory wells often came up dry, but the drillers had expected at least some of these holes in the Arbuckle Mountains to show oil. Buckstaff suggested the oil might have been flushed out by flowing groundwater. That was indeed possible, Hubbert said, explaining he'd already worked out a theory for how it could occur.

After this meeting, Hubbert returned to his earlier ideas on hydro-dynamic traps, gathering his thoughts to present to his colleagues. In internal Shell conferences, he began introducing the notion that ground-water could push oil around. He was a masterful lecturer, with a pro-digious memory. He could speak for hours, filling blackboards with equations, without referring to a single note. He thought it would be straightforward to get Shell to put the idea into action, applying it in exploration. Yet after a year or two of these presentations, he'd made little headway. "What became apparent was that with the exception of a few individuals, pure theory simply wasn't impressing anybody," he later recalled. "Curiously, in our own laboratory, there was a great deal of skepticism of what I was doing, including from the director down."

It reminded Hubbert of situations that Galileo had faced. In Gali-leo's day, during the Renaissance, the accepted authority on the natural world had still been the ancient Greek philosopher Aristotle. Galileo had argued with other scholars over a remark of Aristotle's about a seemingly inconsequential point: that an object's buoyancy depended on its shape, so round objects float, while angular ones sink. Galileo disagreed and tried to reason with the others, with little success. Then Galileo tried another approach, undertaking a long series of experi-ments with balls and blocks and boards of various materials—wood, ebony, brass—which he published in a book titled *Discourse on Float-ing Objects*. Some objects sank, others floated, but it had nothing to do with their shape, Galileo showed, and everything to do with what they were made of. Aristotle was wrong. Galileo also hoped to make a wider point: experiments could refute ideas that had been accepted for centuries.

When it came to the notion of hydrodynamic entrapment, Hubbert figured he'd have to do the same. "I learned this from Galileo," he recalled. "They had to have a demonstration."

A Precarious Position

IN EARLY 1948, IN THE midst of an especially harsh winter, the United States needed more fuel oil than usual—and suddenly ran short. In icebound Chicago, desperate consumers stood in lines holding five-gallon gas cans. Industries shut down across six states to conserve fuel. Frigid temperatures reached as far south as St. Petersburg, Florida, and residents there likewise waited in lines for fuel.

Through the Depression and the war, Americans' consumption had been held in check. But following the war, demand grew explosively, rising so fast that it "amazed governmental and industry experts alike," *The New York Times* reported. Since the start of the war, the number of farm tractors had doubled, and railroads had completed switching their locomotives from coal to diesel. Likewise, homeowners were replacing their furnaces, upgrading from coal to fuel oil, each burning around twenty barrels a year. The typical car consumed gasoline at a similar rate, and the number of cars on the road had soared. "American motorists have gone on the greatest driving spree in history," the president of Esso Standard Oil told Congress. As *Life* magazine summed up the situation, "It was beginning to look as if the US had progressed too fast for the oil industry."

To try to cope with the shortages, the United States cut its oil exports by half. However, by that time the nation barely exported any oil, so this step made little difference. The country was still easily the king of oil, the world's largest producer at 5.5 million barrels a day, as much as all other nations combined. But Americans were also the world's most voracious oil consumers. If the trends continued, the United States would soon arrive at a historic turning point. Consuming more oil than it produced, America would become reliant on foreign oil.

However, the United States could supply plenty of its own fuel, argued Julius Krug, the secretary of the interior, responsible for monitoring the nation's natural resources. Conventional crude oil alone wouldn't supply enough, Krug said, but there was scope for making immense quantities of synthetic oil. The Green River formation underlying Colorado, Wyoming, and Utah, held enormous deposits of "oil shale," a crumbly grayish-black rock that could be cooked until oil seeped out. Similarly, coal could be transformed into gasoline through a technique developed by two German scientists, Fischer and Tropsch, which the Nazis had put into use during the war. The United States had vast coal resources—enough, a Standard Oil researcher estimated, to supply liquid fuels for fifteen hundred years.

Building this new synthetic fuel industry couldn't wait until another war or emergency, Krug warned Congress in early 1948, since it would require "far too much time, materials and manpower." To get it started immediately, he called for a government-funded crash program. If the United States spent about $9 billion—which would pay for, among other things, 16 million tons of steel required for the plants—then within five to ten years, America could ramp up synthetic oil production to 2 million barrels a day, on top of the 5.5 million barrels a day of crude the country produced.

Otherwise, the nation could fall into a trap. "The production center now appears to be moving to the Middle East," Krug argued. "Obviously, as a nation we cannot afford to become dependent on a long supply line from the Middle East for a commodity as important to our well-being as petroleum."

BACK IN 1938, in a brief statement in *Technocracy* magazine, Hubbert had ventured that US oil production would peak by 1950 at the latest. As that date approached, however, he had yet to do an in-depth analysis to put his earlier inkling to the test. During the war, he'd studied where the Allied forces could find resources needed to continue the fight—but he'd been focused on worldwide deposits of minerals like iron, lead, and cobalt rather than on oil resources back home. So far at Shell, he'd been busy trying to improve oil exploration techniques but hadn't stepped back to take in the bigger picture. Since he'd written that 1938 article, US oil production had soared by 50 percent, with no sign of an approaching peak.

Hubbert got an opportunity to revisit this issue in 1948, when he received an invitation to speak at the annual meeting of the American Association for the Advancement of Science (AAAS), an august group. In a session on energy, Hubbert would cover the outlook for fossil fuels alongside two other respected scientists, one covering nuclear power and the other solar power.

Preparing for this talk, Hubbert wrote to a former colleague at the Illinois State Geological Survey, looking for help in finding the latest data. Hubbert explained that on moving to Houston, he'd boxed up all his reference books—and five years later, they still sat in his garage. For the AAAS talk, he'd delve into resource estimates anew, trying to answer questions that had nagged him for more than two decades, ever since his undergraduate economic geology course when he'd first seen graphs of soaring coal and oil production.

AT THE AAAS conference in Washington, DC, held in September 1948 at the Hotel Statler, a few blocks north of the White House, the session "Sources of Energy" led with Hubbert's talk, "Energy from Fossil Fuels."

"Our senses have been dulled by the platitude that 'history repeats itself,'" Hubbert opened. However, the developments that had created the modern world, "far from being normal are among the most abnormal and anomalous in the history of the world." So, he argued, "one cannot refrain from asking: How long can we keep it up? Where is it taking us?"

Populations around the world were climbing at a rate of nearly one percent a year, fast enough to double each century. "That the present rate

of growth cannot long continue," Hubbert said, "is also evident when it is considered that at this rate only 200 more years would be required to reach a population of nearly nine billion—about the maximum number of people the earth can support."*

As living standards increased, consumption of fossil fuels was growing much faster than the population. Fossil fuels were "essentially a fixed storehouse of energy which we are drawing upon at a phenomenal rate," he said. "While the quantities of fuels upon the earth are not known precisely, their order of magnitude is pretty definitely circumscribed." Experts were still arguing over how much of the coal they knew about could actually be pulled from the ground. A 1913 estimate said there was some 8 trillion tons of coal worldwide—but more recently some mining engineers said that perhaps only one-tenth that amount would be practical to recover, since some seams were too thin to mine. Despite such uncertainties, Hubbert argued such figures offered a rough idea of how much coal was available to humanity.

Amounts of petroleum were harder to estimate. Whereas solid coal sits in one place, oil moves through rock wherever there are cracks, faults, and pores. If the migrating petroleum hit an impenetrable "cap rock," it would collect there, forming an oil field. How often this would occur, in various basins around the world, was unclear. Lewis Weeks, Standard Oil's chief geologist, had spent years tackling this problem. His latest estimates held that US land would probably yield, in the long run, about 100 billion barrels of oil. Based on Weeks's work, Hubbert estimated that offshore areas, in particular the Gulf Coast, might hold another 100 billion barrels. "These figures are regarded as being somewhat liberal," Hubbert said, "and the quantity of oil may actually be considerably less." While the quantities were large, petroleum consumption was rising at a phenomenal rate. Over the past decade, he observed, the world had consumed as much oil as it had consumed in all the decades before that.

If crude oil fell short of satisfying the nation's appetite, there were

---

* A world population limit of 9 billion was in the middle of the range of such estimates at the time.

other sources of liquid fuels, Hubbert explained. America could turn to its vast deposits of coal and oil shale, as Interior Secretary Krug had called for. America's oil shale might add another 35 billion barrels, Hubbert said, and Canada's tar sands could yield some 20 billion barrels.

Regardless of the exact amounts, these fossil fuels would go through a life cycle—a rise and fall. On a graph representing the past and future production of any fossil fuel, the curve could take on "an infinity of different shapes," he explained. But such curves all had one thing in common. All would start at zero, rise to one or more peaks, and then decline, falling again to zero.

In this talk, he didn't hazard a prediction. "How soon the decline may set in, it is not possible to say," he stated. "Nevertheless the higher the peak to which the production curve rises, the sooner and sharper will be the decline."

Water power from hydroelectric dams, however, could follow a different pattern, Hubbert said. If more and more dams were installed, hydroelectric power could increase, eventually reaching a plateau, which could persist indefinitely. According to US Geological Survey estimates, hydroelectricity could supply roughly one and a half times more energy than the world was then getting from fossil fuels. So the world's current industrial society could, in theory, run on water power alone. But hydroelectricity wouldn't be able to supply enough energy—no source could—if consumption continued growing indefinitely. Here he showed a graph of various types of growth, including an exponential curve soaring to the heavens, and a bell-shaped curve representing the rise and fall of fossil fuels, and showing how water and solar power could be harvested indefinitely. (The figure was much like one he'd put in the *Technocracy Study Course*.)

In a long-term perspective, the rise and fall of fossil fuels would amount to "but a moment in the total of human history," Hubbert argued. "Our present position is a precarious one." If humanity remained dependent on fossil fuels, when fossil fuels declined, so would modern civilization. They'd be back to living by horse and plow—and the population would collapse.

However, Hubbert argued, if humanity switched to long-lasting sources of energy, such as water power, then a high-energy society could endure for thousands of years. In closing, he struck a note of optimism. "We cannot turn back. Neither can we consolidate our gains and remain where we are," he said. "We have no choice but to proceed, and while our future is unpredictable in detail we can be sure it will be both novel and interesting."

Hubbert had not mentioned two new forms of energy, since the following speakers would cover those. Farrington Daniels, a University of Wisconsin chemist who was a pioneer in solar energy, argued that humanity should "use our rich heritage of sunlight more efficiently," to avoid a horde of future threats—"such catastrophes as war, overpopulation, exhaustion of oil and coal, and return of the glaciers." Closing out the session was Eugene Wigner, a physicist who had been one of the first to realize the potential for an atomic bomb and had helped instigate the Manhattan Project. Wigner was optimistic about nuclear power—but nonetheless cautioned that the world needed to carefully husband its resources.

Hubbert's talk drew little media attention. Most headlines focused instead on Daniels's and Wigner's talks, which mentioned possibilities seemingly pulled from science fiction. Daniels imagined replacing farms with solar-powered factories that would suck carbon dioxide from the air and assemble synthetic foods. A *New York Times* article—headlined "Atom Power for Space Ships to Defy Gravity"—described Wigner's vision of rockets that could escape Earth, reaching the moon and perhaps even hopping to other planets.

Hubbert's scientific peers, though, took his talk seriously. The following winter, *Science*—America's premier scientific journal—printed a version of his presentation, condensed to seven pages. It was the issue's headline piece, announced in bold type on the cover. "That was the first time I was ever swamped with mail for a published paper," Hubbert recalled. "I must have had several hundred letters." He was proud that the responses were "just about one hundred percent favorable."

The Parley

AT A UNITED NATIONS CONFERENCE in New York in August 1949, Hubbert was attending the keynote talk on oil. He had known the speaker, Arville Levorsen, for years and didn't expect to hear anything particularly surprising in this talk. Instead of sitting down on the floor with the other delegates, he was up in the visitors' gallery, taking notes and chatting with a friend.

But in concluding, Levorsen claimed: "Any failure of world supply to meet world demand over the next several hundred years will certainly not be due to a lack of undiscovered reserves, but rather a failure of the discovery effort for one reason or another."

Hubbert thought, "Good God almighty!" He nearly fell out of his seat.

Although Hubbert hadn't issued his own forecasts for world oil, it seemed fanciful to him that there would be plenty for another several hundred years. Following Levorsen's speech, a number of other experts commented that they agreed with Levorsen and were glad to hear his outlook. Hubbert had to get a critical word in.

THE MEETING—OFFICIALLY NAMED the United Nations Scientific Conference on the Conservation and Utilization of Resources—drew

hundreds of researchers from around the world. For three weeks they met on New York's Long Island, in the UN's temporary headquarters, a cavernous building that had housed a gyroscope factory during the Second World War. There they discussed all manner of resources from oil to coal, forests to metals.

The conference was the longtime dream of Gifford Pinchot, a forester and two-time governor of Pennsylvania. He had long argued that fair allocation of resources was crucial for world peace—and toward the end of the Second World War, he seemed to have convinced President Roosevelt of the notion as well. Pinchot, who had technocratic leanings, wanted to gather scientists under the auspices of the United Nations to discuss resources. But before his plan came together, Roosevelt died, then Pinchot too passed away. Independently the United Nations decided to hold a meeting along the lines of what Pinchot had been proposing, and in 1949 the conference finally came together.

As Hubbert wrote at the time, the importance of the conference was "difficult to overestimate." It was the first to gather so many scientists to discuss crucial issues facing humanity, and, he argued, it marked "a milestone in the advance of science itself in that it explicitly presents the problems of human society as a legitimate domain for scientific enquiry."

For the keynote on oil, Arville Levorsen was a natural choice. A Stanford University professor who'd previously worked in the oil industry for many years, Levorsen was known for his skill in making complex problems easy to grasp. And, as an American, he represented the nation that produced and consumed more oil than the rest of the world put together.

Levorsen's talk, titled "Estimates of Undiscovered Petroleum Reserves," opened with a review of the methods geologists used to make such estimates. He cited work by a few veteran oilmen, including Weeks of Standard Oil (whom Hubbert had also drawn on as his main source for such resource figures). By Levorsen's tally, these estimates arrived at an "impressive figure" of 1,500 billion barrels of oil remaining to be found around the world—roughly 500 times larger than humanity's yearly consumption at that time.

Levorsen pointed out that geologists had suffered recurring fears—in the 1920s and again in the 1930s—that oil supplies were about to run short. And, of course, such fears had come to the fore again in 1948. But there was no reason to worry, he argued. Even his enormous figure of 1,500 billion barrels was likely an underestimate. Repeatedly, the best estimates at any moment had, in later years, been proven too low. The latest estimates, Levorsen argued, could turn out to be "but a fraction" of the estimates to come in future decades. So, Levorsen surmised, the world should have plenty of oil for "several hundred years."

After Levorsen finished, the head of the Geological Survey of Great Britain commented that he was old enough to remember how the industry had suffered "periodical nervous attacks about the amount of oil available." However, he pointed out, the Middle East turned out to have vast resources, much higher than initial estimates. Other experts were likewise cheery. The chief of Canada's Mineral Resources Division said, "I am sure we are all grateful, all of us who have to drive cars or who have to get around New York City, for the very optimistic spirit of Dr. Levorsen's paper."

Hubbert immediately left the visitors' galley and made his way toward to the delegates' floor, jittery with nerves, determined to challenge these optimistic views. When the chairman opened the session to questions from the floor, Hubbert spoke up.

He began boldly, calling Levorsen's reasoning "an exercise in meta- physics." He conceded that estimates were never perfect, that it was true that in the past geologists had been too pessimistic about oil resources. However, by the mid-twentieth century, geologists had developed "per- fectly objective estimates," he argued, which gave them a rough idea of how much oil remained. Even if the latest estimates were still too low, geologists knew the world had a finite amount of oil. The estimates couldn't keep going up and up forever.

Hubbert also argued that Levorsen's way of framing the issue was misleading. During the first half of the twentieth century, oil consump- tion had soared. Given the world's seemingly unquenchable thirst for oil, consumption would presumably continue to rise. So what seemed at the moment like a five-hundred-year supply would actually get used up considerably sooner.

Also, long before the oil ran out, production would hit a peak, then enter a long decline. "If it continues to go up at anything like the present rate of spectacular increase," Hubbert told the conference, "it can come down just as spectacularly in a much shorter period of time."

Levorsen responded by sticking with his argument, maintaining there was far more oil to discover, with no limits in sight. How much would Earth ultimately yield? As yet, Levorsen told the audience, "no one knows."

THE NEXT MORNING, the front page of *The New York Times* carried the headline, "Estimate of 500-Year Oil Supply Draws Criticism in U.N. Parley," and laid out the quarrel between Hubbert and Levorsen. A few days later the newspaper ran an editorial that pointed out how surprising this rift was. "Oil companies usually take a cheerful view of the oil future," the editorial said. "Dr. King Hubbert departed from their custom when he branded Professor Levorsen's figures as 'an exercise in metaphysics.'" However, like the experts who'd been invited to comment at the conference, the *Times* sided with Levorsen, who in its estimation was "to be taken seriously." The editorial was titled "More Oil Than We Think."

This debate made Levorsen "decidedly embarrassed and irritated," Hubbert recalled, and led to "a ruckus among the oil people" at the conference. These participants organized an impromptu seminar to review oil estimates, which drew some forty or fifty participants—including the director of the US Geological Survey, experts from the US Bureau of Mines, and others from academia and industry. Some argued Levorsen's estimates were based on fragmentary evidence and might be far from correct. Others shared Levorsen's optimism. The director of the Geological Survey of Great Britain reiterated the sunny view he'd presented during the main session on oil, saying there was enough oil already known to last 150 years at the current rate of consumption. Ira Cram, vice president of

Continental Oil, also strongly supported Levorsen. "The philosophy that we will find more oil as we learn more about the earth's crust," Cram argued, "is true, and is going to be true for a long time." In the end, the seminar resolved nothing. The consensus, according to a reporter who sat in on the talks, was that "nobody knows how much petroleum there is inside the earth, or how many years or hundred years it will last, and nobody is likely to know soon."

This attitude frustrated Hubbert. The optimists were arguing that no one could accurately gauge how much oil there was—yet also gave reassurances that there would be plenty long into the future. They seemed to take comfort in the ignorance of generations past, and to hold faith that future generations would prove their own best estimates to be paltry. Yet they didn't seem to grasp how quickly the oil would be consumed, if consumption continued growing exponentially.

Hubbert felt his efforts weren't a total loss, though. "If I hadn't done it," he later said, "I figure Levorsen would have gone on the record without any discussion, as the official opinion of the meeting. I broke it up."

ONLY AFTER THE three-week UN meeting was over did Hubbert finally find time to gather his thoughts about his clash with Levorsen. Forgoing a vacation he'd planned, over the following week he wrote up a more detailed, systematic critique, in which he laid out his own thoughts more clearly—both for his superiors at Shell and for the journal *Geophysics*, which he'd been editing for the past couple of years.

Hubbert agreed that Levorsen's estimate—that there was about another 1,500 billion barrels remaining to be found worldwide—was reputable. But he dismissed Levorsen's prediction that the industry might wind up discovering several times more than that. Instead, he took the estimate at face value. To calculate a figure for the ultimate amount of oil that the world might extract, past and future, he used Levorsen's estimate for undiscovered oil and added to that the amount already extracted, as well as the "proved reserves"—the quantities likely to be profitable to extract from fields already discovered. In this way, Hubbert arrived at a value of about 1,600 billion barrels. The world could burn through this quantity fast, or burn through it slow, but either way the production rate was likely to reach a peak and then decline. If that peak production rate were about twice the current rate, Hubbert wrote, the world might reach that peak in about seventy-five years—around the year 2025. Or if the world burned oil faster, then the peak could come in fifty or sixty years—in the early 2000s. To get the idea across, he drew two curves, one taller and skinnier, the other shorter and wider.

Looking at it this way, Levorsen's supposed five-hundred-year supply appeared sufficient for only a few generations. "Once this peak is passed," Hubbert concluded, "the industry will manifestly not be able to meet world demands."

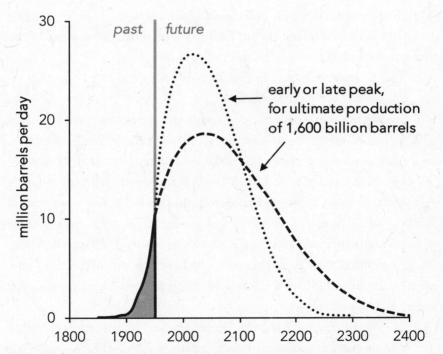

World oil production: Assuming the amount of oil available worldwide would ultimately be about 1,600 billion barrels, Hubbert estimated production would likely peak during the first half of the twenty-first century.

. . .

WHILE ON HIS trip to New York for the UN meeting, just before head-ing back to Houston, Hubbert stopped to meet with Technocracy. The organization still had its "Continental Headquarters" in Manhattan, on the twenty-fifth floor of the Commerce Building, occupying an eight-room suite. It had been more than fifteen years since Hubbert helped found the organization, and since moving to Houston, he had drifted away from it. He was still on Technocracy's board of governors but rarely attended its meetings.

It may have been on this trip that he had a confrontation that other members of Technocracy would recount decades later. When Hubbert arrived for the meeting, he was visibly drunk. It was the first time some of the core Technocrats, such as Scott's longtime secretary, had ever seen Hubbert in such a state. He was clearly upset.

Hubbert asked one of them, "Scott did give a date for the end of the price system, didn't he? Didn't he predict that?"

"No, he never has," the Technocrat replied.

"You sure?" Hubbert replied.

Ever since its early days, the organization had claimed it had a remark-able record of predicting the future. A 1938 article in *The Technocrat*, a magazine of the organization's Los Angeles section, asserted, "Technoc-racy Inc. has been able to accurately predict trends and events. Its strength lies in the accuracy of its predictions; Technocracy Inc. has never missed a major one." For years, Scott had been predicting North America's price system would collapse and that it would become a Technate. In 1935, Scott suggested that Technocracy would have enough members to take over by 1940. In talks in 1937 and 1938, he predicted a collapse of the price system by 1942.

During the war, the nation had moved away from running a relatively free-market "money economy" toward a managed "goods economy"—but had never approached anything close to what Technocracy envisioned. The New Deal's deficit spending had kept the economy from collapsing through the 1930s, then military spending during the Second World War had revived the moribund economy—and the boom continued after the war, driven in part by the burgeoning military-industrial complex. By the close of the 1940s, it was abundantly clear that the "price system" had weathered the Great Depression and the world's greatest war—and was still as ensconced as ever. The nation remained firmly capitalist. There had been no collapse.

After that 1949 board of governors meeting, Hubbert never attended another Technocracy meeting.

THE FOLLOWING YEAR Hubbert had another major change in his life. Over the past decade, he'd periodically looked into the marriage laws in various states, but he and Miriam never had wed. In August 1950, after fifteen years together, when both of them were in their mid-forties, they went to Las Cruces, New Mexico, and finally, officially, got hitched.

Then in 1951 Hubbert got a promotion of sorts at Shell. He'd been the assistant director of the Bellaire lab for several years but had always been an uncomfortable presence there. The lab's employees prided themselves on speaking plainly to each other about ideas—but Hubbert took this farther than his colleagues. Many of them had stories about Hubbert's intellectual takedowns. Once, when some researchers gave a presentation on the movement of fluid through a tube full of glass beads, they said it had a certain low permeability, meaning fluid wouldn't easily flow through it. Hubbert, from the back of the room, blurted out, "That glass bead pack has a permeability of several Darcys"—far higher than the researchers had calculated. "You're not doing the experiment you thought you were doing." With performances like this, a saying caught on around the lab: "That Hubbert is a bastard, but at least he's *our* bastard."

Given Hubbert's stubborn, irascible ways, Shell created a unique position for him. In many ways it was like the job as technical adviser that Shell had talked about giving him on hiring him years earlier. He'd have no assigned duties and wouldn't need to manage others. He'd be free to pursue his own interests and would help various teams solve problems. He got to pick his own title, and settled on "Chief Consultant, General Geology."

Stampede

AFTER SEVERAL YEARS OF HUBBERT'S attempts to convince his colleagues at Shell to test his idea about hydrodynamic traps—the notion that flowing groundwater could push oil into unexpected places—in 1951 a young geologist named Gilman Hill wrote a paper on the same concept. Hill had studied Hubbert's 1940 paper, "The Theory of Ground-Water Motion," and, intending to apply it to oil, had gone to Stanford for a year to study under the geologist Arville Levorsen.

"The theory was valid," Hubbert recalled, "but it didn't contain anything that wasn't contained in my previously published papers, which he admitted he'd used and studied." Where Hill's paper covered ground-water flows, Hill's equations were almost identical to Hubbert's, showing similar stylistic idiosyncrasies, Hubbert thought, and he considered one of Hill's illustrations as essentially the same as one of his own. Soon afterward Hill and Levorsen went into business together to search for oil in hydrodynamic traps—an attempt to "grab off an oil field or two before anybody else caught onto it" was Hubbert's assessment. Whatever the merits or deficiencies of Hill's paper, others were onto Hubbert's idea of hydrodynamic traps before Shell had even put it to the test.

"I wrote a very hot letter to the management," Hubbert recalled. "I simply demanded the privilege of writing a paper on my work on this thing." Shell usually didn't let its researchers publish work that could give rivals an advantage in exploration. But with the idea of hydrodynamic traps already out there, Shell gave its consent. Shell then finally drilled its first test well in search of a hydrodynamic trap, in Wyoming's Big Horn Basin—but after drilling 2,506 feet, abandoned it as a dry hole.

To present his theory to a wider audience, Hubbert lined up a slot at the annual conference of the American Association of Petroleum Geologists (AAPG), held in Los Angeles in March 1952. His friends put out the word that Hubbert's lecture would be a must-see. When his talk came around, at the Biltmore Hotel, its biggest ballroom was packed.

Over the decades, Hubbert told the audience, ideas of how oil got trapped underground had undergone an evolution. In the early years of the twentieth century, geologists had come across oil in all manner of geological structures—sometimes with a good reason for drilling where they did, other times just by luck. In the 1920s, some had promoted the "hydraulic theory" that water could push oil around underground, but that idea had remained vague. Over the years, the hydraulic theory had been sidelined. A consensus had formed, still in place by the mid-twentieth century, that oil could be trapped only by a barrier of rock, and the industry ignored any effects from flowing groundwater. However, "not only are these effects theoretically expectable," Hubbert argued, "but they have been found to occur in almost every major oil-producing area." If groundwater wasn't flowing enough to affect oil fields, then the layer of oil trapped in the rock would be horizontal. Yet there were many fields where the oil was known to sit at an angle, a situation that could be maintained only if constantly buffeted by flowing groundwater.

Hubbert went further, arguing that groundwater flows could push oil into unexpected places, outside the types of rock formations that the industry usually searched for. So, he declared, geologists needed to take groundwater into account "if many such accumulations are not to be overlooked."

As soon as Hubbert finished his talk, many in the audience rushed out. "There was a stampede for the telephone booths to call their home offices," as he put it. The notion of hydrodynamic traps was evidently still news to many in the industry.

Hubbert's talk was so successful that the AAPG invited him to go on tour as a distinguished lecturer, as they had a decade earlier with his talk on laws of scaling. The tour on hydrodynamic entrapment was far more extensive, with some thirty stops across the United States and Canada. Meanwhile he finished writing up his ideas for the *Bulletin of the AAPG*. His paper, "Entrapment of Petroleum Under Hydrodynamic Conditions," stretched to nearly seventy-five pages, about half of it mathematics. (According to Marlan Downey, who later served as Shell's chief geologist, that study "must have set the AAPG record for number of pages containing equations.") It was, Hubbert told a fellow geologist, "potentially one of the most important papers I have ever written."

On publication, Hubbert's work quickly won accolades—including a kind letter from Wallace Pratt, an oil industry veteran dubbed the "geologist's geologist." Then nearly seventy, Pratt had earned his reputation during decades with Humble Oil, using his deep understanding of the earth to uncover huge deposits in areas others had written off, from Oklahoma to Mexico. Pratt had then worked his way up to top management positions, reaching vice president of Standard Oil, finally retiring at the end of World War II. Over the previous few years, Hubbert and Pratt had exchanged a few letters. After reading Hubbert's entrapment paper, Pratt wrote to him:

> It explains many facts that have long puzzled me. I could cite occurrences out of my long past experience which now seem to me to be illustrations of accumulations of petroleum under the influence of hydrodynamic conditions. When we first observed this type of accumulation we were at a loss to explain it, but repeatedly we made successful locations [of oil] by staying off the crest. I have in mind particularly our experience in North Central Texas in 1916.

In the following months, Hubbert's work gained praise from many quarters. In the fall of 1954, when the Geological Society of America gave Hubbert its prestigious Day Medal for his work in geophysics, it cited his many accomplishments, including the "spectacular results" in his entrapment paper. And Arville Levorsen—despite his disagreements with Hubbert over oil resources—called his entrapment paper "a classic."

Shell made a few more efforts to put Hubbert's idea into practice. A second test well in search of a hydrodynamic trap found a bit of oil but not enough to be a commercial success. A third attempt found a field, holding an estimated 18 million barrels—just enough to be profitable to develop. The fourth test well came up dry. Hubbert blamed the disappointing results on bad luck. In a report to management, he wrote, "It has been purely fortuitous that the Shell wells were dry holes, or made only marginal discoveries, when competitors were having better luck in finding larger hydrodynamic accumulations," including two Wyoming fields known as Northwest Lake Creek and Murphy Dome.

Meanwhile some of Hubbert's colleagues at Shell argued his idea wasn't much use. For decades, explorers had been finding fields that, in retrospect, must have been hydrodynamic traps. Such discoveries puzzled them, as they had Pratt. But even if they didn't have any strong reason for "staying off the crest" of anticlines, for example, they'd figured out rules of thumb that allowed them to find more such oil fields. In light of this record, some of Hubbert's colleagues argued, they didn't need to explicitly consider flowing groundwater.

Their reasoning astounded Hubbert. Shell hadn't put out much effort to look for hydrodynamic traps. There could be many such fields left to find, he figured—and they'd be more successful in finding such fields if they understood the underlying factors. Hubbert was worried about the industry's ability to find enough new oil fields, but his colleagues did not seem to be.

A Magical Effect

AT A 1954 CONFERENCE, TWO oil companies—Stanolind on one side, Atlantic Refining on the other—got into a fierce debate about the physics behind hydraulic fracturing. This relatively new method involved pumping fluid into a well until it opened fractures in the rock surrounding the well, yielding more oil. The two sides "got a dogfight going that lasted nearly all afternoon," Hubbert recalled, laughing. Neither side's explanation made sense to him. Nonetheless he stayed "on the sidelines just kind of enjoying the show."

Stanolind had introduced the new technique in 1948, under the name "the hydrafrac process"—soon shortened to simply "fracking." The technique had been an immediate hit. In just its first several years, it was applied more than thirty thousand times, helping extract more oil, especially from older wells.

However, even after its use had become widespread, no one knew exactly how hydraulic fracturing worked. Since it was put to use at the bottom of wells, thousands of feet underground, there was no easy way to see the results of the technique. Why did the rock fracture at a particular pressure? And why did the required pressure vary so much from well to well and place to place? No one was sure.

The debate Hubbert observed at the 1954 conference was over a seemingly esoteric point, but it was central to understanding how the fracking process worked. Stanolind argued the fractures were horizontal, like pancakes. Atlantic said the fractures were vertical, along the length of the well, like the fins on a rocket. Each company had its own reasoning—and Hubbert thought they were both wrong.

Soon after that contentious conference, Shell gave Hubbert a new assignment: figure out how fracking worked.

HUBBERT HAD FIRST encountered hydraulic fracturing in 1946—although rather than trying to put it to use, it was a problem Shell wanted to avoid. The company's engineers had come to Hubbert to see if he could solve a mystery that had arisen as they drilled offshore in the Gulf of Mexico. As with most every oil well, these engineers were using a concoction called drilling mud—a slurry of clay and other minerals. They pumped it down the hole, where it flowed around the drill bit, cooling it, and also clearing out chips of rock from the bottom of the hole.

The mud served an additional, particularly crucial purpose. As it flowed back up the hole, the mud put pressure on the walls of the well, holding back oil and gas under high pressure, which otherwise might unexpectedly blow out. In decades past, such gushers—the iconic black fountains erupting from oil derricks—had been a cause for celebration. But they were wasteful and dangerous, and so had been mostly tamed through careful use of drilling mud.

The Shell engineers told Hubbert how, in drilling along the Gulf Coast, they'd gone down some six thousand feet using very light, watery mud without encountering any problems. But in a span of one or two hundred feet, they'd encountered a zone with much higher pressure that required heavier mud to supply enough pressure to hold back the fluids in the rock. They'd worried that if they waited too long to use heavy mud, they'd risk a blowout. Yet when they put in the heavy mud too early, they suffered another problem, known as "lost circulation," in which the drilling mud simply disappeared down the hole. This lost circulation was a major problem for drilling in this area, slowing down the process and driving up costs. Some wells suffered lost circulation so chronically that they were abandoned before they struck oil. Not sure of the dynamics underground, the Shell engineers couldn't determine how to prevent lost circulation.

Hubbert's first thought was that they must have hit a limestone layer that had cavernous holes in it. But the engineers told him there was no limestone in that area. Then they explained there was another aspect to the mystery. When they sent down mud at high pressure, the mud disappeared. But when they eased off the pressure, the circulation returned. The engineers and Hubbert agreed that the pressure from the heavy mud must be opening up fractures surrounding the well.

However, the pressure wasn't enough to lift the overburden—the weight of all the rock pressing down from above—and that puzzled the engineers. Hubbert realized the engineers were imagining the fractures as pancake-like horizontal openings in the rock—and for such cracks to open, they'd have to lift the weight of all the rock above. Hubbert told them it was far more likely that the fractures were vertical, so to open up they'd only have to push neighboring rock sideways, which would usually require much less force. It was like the difference between trying to lift the heavy door of a bank vault off its hinges, compared with swinging the door on its hinges.

Hubbert had solved the problem. His explanation of the underlying physics allowed the engineers to adjust their drilling mud to avoid lost circulation and drill more efficiently.

MEANWHILE STANOLIND HAD also discovered hydraulic fracturing and devised a method for putting it to use. At first it had been experimental, using a fluid that was cheap to buy. "Due to availability and price, war-surplus Napalm has been used in the majority of experiments to date," Stanolind reported in 1948. They mixed sand into the napalm— also known as "jellied gasoline"—which carried the sand along with it, enabling the sand to lodge deep within the newly opened fractures. Then they pumped down another fluid that dissolved the jelly, and much of the napalm flowed back to the surface. But the grains of sand remained stuck in the fractures, propping them open, allowing oil and gas to escape from the rock.

The technique quickly caught on, and companies advertised their fracturing services and equipment in industry magazines like *Oil and Gas Journal* and *World Oil*, commonly showing drawings of horizontal, pancake-like fractures. Stanolind had based this on evidence drawn from a few shallow test wells and had assumed that all fractures, even those deep down, would likewise be horizontal.

This explanation was "accepted almost universally, with rarely a dissenting voice," as Hubbert saw it. But to him, that explanation made no sense. They'd got the physics wrong. In 1953 he had published a short commentary in the journal *Petroleum Transactions*, arguing that Stanolind's explanation was mistaken and the fractures were likely vertical. At the 1954 conference where the "dogfight" broke out over fracking, both sides had explanations that differed from Hubbert's. Although the technique was in widespread use, there was still disagreement over how it worked, and its application remained hit or miss. There was clear room for improvement.

By the time Shell tasked Hubbert with explaining how fracking worked, "we had the records of several thousand fracturing jobs, with varying degrees of reliability in their data," he recalled. "We had to smoke out useful information."

Hubbert had recently hired a new assistant, David Willis, who had just completed his doctorate in geology at Stanford. In the fall of 1954, Hubbert and Willis developed a theory that described the forces at work when fluid was pumped into a well at high pressures. They worked on this intensively for a few months, writing up the results for a high-level conference to be held the following summer at Royal Dutch Shell's lab in Amsterdam. Hubbert and Willis found that in some special conditions— in shallow wells, or along faults under sideways stress, like California's San Andreas Fault—the fractures would lie horizontal. But their analysis suggested that in most cases the fractures would be vertical, along the well's shaft.

When Hubbert presented the results at the Amsterdam conference, "it was very well received by the highest level technical people," he recalled, "accepted completely, with no significant criticism." After this vote of confidence, Shell organized training sessions on the new analysis for its field engineers. When it came time for the first course, Willis was away so Hubbert gave it himself. "What I discovered was that the theoretical argument was having no effect whatever on these men," Hubbert recalled. The engineers were absolutely sure that the fractures were horizontal. Every article, every ad on fracking showed fractures oriented that way. They had been "completely brainwashed," Hubbert thought. "They didn't have any real evidence, but they'd been so thoroughly indoctrinated on this thing that they knew damned well these fractures were horizontal." It mattered, because if they didn't understand the forces at work, they couldn't control it precisely. The technique would remain more art than science.

When Willis returned to Houston, Hubbert told him the presentation had been a flop. Willis didn't say much at the time. But a few days later, on a Monday morning, Willis appeared in the doorway to Hubbert's office, looking anxious. He wanted Hubbert to come to the lab to see something. Swamped with backlogged paperwork, Hubbert told Willis it would have to wait. Willis left, then came back half an hour later, getting more and more fidgety. He'd been working on something over the weekend, he said, and Hubbert should come and see it. Hubbert relented and trudged over to the lab.

In Shell's Bellaire lab, one of the nation's best-funded research facilities, sat the contraption Willis had assembled at home over the weekend. It was a goldfish bowl, filled with liquid Knox gelatin and some plaster in it. Willis had used the gelatin to simulate rock—appropriate, given Hubbert's work on laws of scaling—and had stuck an Alka-Seltzer bottle in the middle of it to mimic a well. He'd put the liquid gelatin in the fridge and let it set, then pulled out the bottle. Then he'd used a baster to pump a slurry of plaster of Paris down the hole, filling it until the plaster began to push its way into the gelatin, forming fractures. As their theory predicted, the fractures were vertical.

Although Willis's setup was kludged together, Hubbert immediately realized it was what they needed to win over the field engineers: a clear demonstration. They'd have an opportunity to make their case at an internal Shell conference in early 1956, in several weeks' time. They got to work on building a larger version of the model. To replace Willis's goldfish bowl, Hubbert scoped out bigger aquariums on sale at local shops.

At the Shell conference, Hubbert and Willis explained their experiment and showed the plaster casts, first from one angle, with the fractures flaring out from either side of the well. Then they rotated the cast, so the audience could see that the fractures were thin and sharp, like a knife's blade. And of course, they were undoubtedly vertical.

Within a week of this demonstration, field engineers began sending in data they'd collected after fracturing wells. Some of them had put rubber plugs down wells to form an impression of the wall. Others sent cameras down the hole. This field data showed the fractures were indeed vertical. The theory was right—and finally the engineers believed it. Willis's contraption "had a magical effect," as Hubbert put it. "It made Christians out of these people."

Sweeping Under the Rug

EVEN AS HUBBERT BUSIED HIMSELF explaining the physics of fracking, he got pulled into an area new to him: the secretive realm of nuclear power. Through the Second World War and afterward under President Harry Truman, nuclear energy in all forms—both weapons and power—had been highly classified. President Dwight Eisenhower took a different approach. In a 1953 speech before the United Nations, he had called for the world's nuclear powers to open up access to the technology and supply more nations with reactors as well as the uranium to power them. The approach was dubbed "Atoms for Peace."

As part of this new openness, the Atomic Energy Commission (AEC)—the US agency tasked with encouraging nuclear power as well as regulating it—sought input from the scientific community. In the fall of 1954, the AEC asked the National Research Council (NRC) for advice on disposing of nuclear waste. With the first commercial nuclear power plants in the works and the promise of many more, the nation would generate ever larger volumes of radioactive waste, which would have to be dealt with somehow.

To make recommendations on safe disposal of these wastes, the NRC engaged a handful of top scientists with the right kinds of expertise, including Hubbert. The vast majority of the waste would be liquids left over from refining uranium and other steps involved in preparing and running a nuclear reactor, and most ideas for disposing of it involved sticking it underground in some fashion. So there were a number of fields in which Hubbert had relevant expertise: on faults, on the strength of rocks, and on fluid flow. By then, Hubbert's work on scale models was considered a classic, and his hydrology study was becoming more widely accepted and cited, so he was a natural choice for the panel.

In November, the NRC's panel of scientists met in Washington DC, in one of the small temporary buildings on the Mall. The building had a level of security Hubbert considered ridiculous. "We were just about everything but fingerprinted to get in the place," he recalled. The visiting scientists were supposed to have a minder with them at all times. "I got arrested," he griped, "for trying to go to the gents room without an escort."

During the day-long meeting, the AEC ran through masses of information about nuclear wastes. When uranium atoms fission, they split apart, releasing radiation and creating a stew of other elements. Many of the elements involved came in slight variants known as isotopes, some of which were unstable and radioactive, others stable and benign. As a nuclear neophyte, Hubbert found the presentation difficult to follow. Late in the day, things opened up for discussion. The chairman of the session told the gathered scientists, "All right now, what we want you to do is tell us what to do with this stuff."

"I've sat here all morning," Hubbert spoke up, "and up until now I've been trying to get an answer to a couple of questions that it seems to me we need to know. Maybe you've told us but if so I missed it. Approximately how much of this stuff per year are you producing? And approximately what are its physical properties?"

The chairman looked around. That information was classified, he replied. Hubbert got no answer. "The whole thing was ridiculous," Hubbert thought. "Here was the very information we had to have, and that was secret."

In any case, the AEC apparently considered this initial meeting a success. It soon requested an in-depth study from the NRC, which invited Hubbert to join this more permanent effort, which would be called simply the Committee on Waste Disposal.

SEVERAL MONTHS LATER, in the spring of 1955, Hubbert got good news from his friend James Gilluly, a geologist with the US Geological Survey and a member of the National Academy of Sciences. Hubbert was in the running for election to that elite group, Gilluly said, and urged Hubbert to come to Washington to the luncheon where the new members would be announced. Hubbert mulled it over for "several restless nights," he wrote to Gilluly, but decided against attending, thinking it would be "neither dignified nor fittin' for me to be standing by at that particular moment."

In April, Hubbert got word from the academy that he had indeed been elected a member—one of the highest honors a scientist could receive. To share the good news, he wrote to a few close friends. He also wrote letters to his boss at Shell and to the company's president, H. S. M. Burns, in which he defended his combative approach:

> I seem to have an aptitude for arriving at conclusions which are contrary to the "authoritative" opinions of the moment. To me, such things are worth fighting for despite the immediate unpopularity that almost inevitably results. I interpret my election to the Academy in the face of such a record as being a vindication of this attitude.

In these battles, Hubbert felt he was not simply defending his own ideas. In a letter of thanks to Gilluly, he said, "As you well know, I feel deeply about scientific enquiry, which, it seems to me, is the only thing standing between us and the Dark Ages."

. . .

AS HUBBERT DELVED deeper into work for the nuclear waste committee, he reconsidered his long-term outlook for energy. He'd been wondering for decades: How would humanity ever replace fossil fuels? For the past quarter century, he'd expected that hydroelectric power would be the only large-scale, long-lasting energy source.

At the time, Hubbert wasn't impressed with nuclear power, because the fuels seemed so rare. "I was very skeptical that uranium could ever amount to anything as a source of power," he recalled. "Atom bombs, yes—they had enough atom bombs to blow us off the earth. But it was not very promising for power." Nonetheless, he wanted to find out what he could about nuclear power's potential.

The AEC still kept the vast majority of information on nuclear power classed as "confidential" or "secret," and still tightly controlled operations at its sites scattered across the country, from the uranium enrichment plants at Oak Ridge in Tennessee to the plutonium plants at Hanford in Washington state. Nonetheless, information on nuclear power was becoming more readily available, as part of Eisenhower's Atoms for Peace drive. The AEC declassified thousands of documents, and in August 1955, at an international scientific meeting in Geneva, many countries shared, for the first time, details on nuclear enrichment and reactors.

A few weeks after the Geneva conference, the waste committee had its first formal meeting, where it brainstormed with the AEC. In opening, an AEC staff scientist, Arthur Gorman, made a frank admission about the agency's handling of nuclear waste. "To some extent, because of our geographically isolated locations, it has been possible to 'sweep the problem under the rug,' so to speak," Gorman said. "But those of us who are close to it are convinced that we must face up to the fact that we are confronted by a real problem."

They then heard from a variety of scientists, from both the AEC and the Geological Survey, who described the way nuclear wastes were being disposed of. Once fuel rods of enriched uranium were spent, no longer able to provide enough power, they were reprocessed to recover the remaining uranium—so they weren't considered waste. What was left over from enrichment and reprocessing, predominantly liquids, made up most of the waste and ranged in radioactivity from "high level" to "low level."

At the AEC site in Hanford, much of the low-level waste was getting poured into holes in the ground, where the liquid flowed through dirt. The researchers argued that most of the radioactive elements would stick to the dirt, effectively filtering the liquids and leaving them relatively clean. However, the researchers monitoring the waste disposal couldn't account for how much of the radioactive material was actually sticking to the dirt. Nor did they know where all the liquid had gone. A Geological Survey researcher gave a presentation on the situation at Hanford, saying the liquid wastes were either "trapped in the underlying area or have gone down somewhere." Others at the workshop repeatedly asked for more information, saying, "What happened to the rest of it? Where did it go?" Nobody knew.

Other speakers at the workshop described practices at Oak Ridge National Laboratory, where the lab had dug large pits out of shale rock and was running low-level wastes into them. The idea was that the shale would filter the radioactive material out of the fluid. But there, too, they knew little about where the liquids were going after they poured them in, or how effective the filtering might be.

Liquid wastes with intermediate levels of radioactivity were being stored in steel tanks—but how long those tanks would last was an open question, since the nuclear waste was highly corrosive. Storage in these tanks was meant to be only a temporary measure. But there was little effort to develop a more comprehensive way of handling the wastes.

Despite these issues around waste disposal, Hubbert was optimistic they could be solved. With the information he got through the committee about uranium resources, and how much energy was released in fissioning, he changed his mind about the potential of nuclear power. He realized, as he put it, "a little bit of uranium had a hell of a lot of energy."

. . .

TO SEE AEC'S work firsthand, the Committee on Waste Disposal arranged to visit a couple of the commission's sites that were both cradle and grave of the nation's nuclear wastes. Despite moves to open up nuclear technologies, access to these sites was still restricted and the details of their work classified. To go on the tour, each member of the waste committee would have to get "Q clearance"—one of the highest security levels, similar to the Department of Defense's Top Secret. However, the meeting minutes recorded, "Dr. M. King Hubbert prefers to be barred from these field trips because he does not want the responsibilities attached to the Q clearance requisite for visiting the AEC installations."

Only a year earlier, Senator Joseph McCarthy's anti-Communist witch hunt culminated with televised hearings in which he went after the US Army, even accusing George C. Marshall, a five-star general, of being a traitor. And it had likewise been only a year since physicist Robert Oppenheimer—former head of the lab in Los Alamos that had developed the first atomic bombs—was put under investigation and stripped of his security clearance because of ties he'd had with some Communists and "fellow travelers" in the 1930s.

So it may not have been the responsibilities of Q clearance that Hubbert objected to but the extensive background check. Although he'd ceased all involvement with Technocracy, his past role would be easy for anyone to uncover.

19

Jolted

TOWARD THE END OF 1955, Hubbert got an invitation to speak unlike any he'd received before. This latest talk would be at a meeting of the American Petroleum Institute (API), the industry's main lobbying arm. The API took the eternally optimistic view that the nation would have plenty of oil "for the foreseeable future." The API asked Hubbert to speak about oil, both US and world supplies—but the exact focus was up to him. He decided that his talk would cover the history of all the fossil fuels—oil and natural gas, of course, as well as coal. Also, he told the API, he'd take "a critical look at the possibilities for the future."

Reading up on oil resources, Hubbert tried to assimilate "practically everything in print," as he put it, and also canvassed his colleagues at Shell, to make sure his numbers reflected the latest information. In early February, two months into his preparations for the San Antonio talk, Hubbert received a letter from his former boss, Harold Gershinowitz, who'd moved up from director of the Bellaire lab to become president of Shell Development, overseeing all of the company's research. Gershinowitz explained that Shell would soon launch a new training program, aimed at cultivating the company's next generation of executives. To prepare these men to steer one of the world's biggest oil companies, they would hear a sweeping lecture, "Energy Resources of the World." Gershinowitz had already told the course organizers that Hubbert was the right man for the job. Would he do it?

Although Hubbert was extremely busy, this was also an opportunity to shape the views of the industry's future leaders. He replied he'd be honored to give the lecture. It happened that, in preparing his API talk, he was in the midst of "a reappraisal of the whole energy picture," he told Gershinowitz. Like a chemist hunched over test tubes, trying to isolate some pure substance, Hubbert had been boiling down a century of data and forecasts, aiming to "reduce it to its elemental simplicities."

In reviewing how much oil might be ultimately available, Hubbert still thought the best estimates were from Lewis Weeks of Standard Oil. But for the United States, Hubbert tweaked these numbers slightly, rounding up some figures to arrive at a grand total of 150 billion barrels —still somewhat lower than the 200 billion barrels he'd stated as a very rough estimate in his 1948 talk, "Energy from Fossil Fuels." For world oil, he thought that earlier figures he'd used were a bit too high, and more recent work put the total at around 1,250 billion barrels. Although these numbers were rough, they were likely close enough to illustrate the implications.

When he'd drawn a couple of different scenarios for the future of global oil, in his critique for the United Nations meeting several years earlier, there was a lot of wiggle room. The world had used up only a small fraction of the estimated ultimate amount of oil, which meant the future was highly uncertain. People could burn through the remaining oil quickly or they could consume it more slowly.

But when it came to US oil production, given the constraints he'd set up, he found he had little leeway in drawing the curves. If production followed a roughly bell-shaped curve, the peak would have to come awfully soon.

ON THE AFTERNOON of March 6, 1956, Hubbert and his wife began the two-hundred-mile drive, their car loaded down with seven hundred copies of his long paper, plenty for everyone at the meeting to get one. They headed for the swanky Plaza Hotel, a downtown tower beside canals where stone footbridges arched over the water and gondoliers poled past tropical trees—a tourist trap that San Antonio billed as "The Venice of Texas."

Around five p.m. Hubbert and his wife pulled up at the Plaza's entrance to unload hundreds of copies of his report. He was surprised to find himself suddenly surrounded by journalists. A variety of publications—the *Oil Daily* and *Petroleum Week*, the Associated Press, and the *San Antonio Express*—were there to cover the meeting. These journalists all wanted to know one thing: Did Hubbert still plan to deliver his speech?

The API had issued a press release that opened with Hubbert's stark forecast: "Production of oil and gas in the United States will reach its peak about 1965 and will decline thereafter at a rate comparable to the preceding rate of increase." This outlook was, it seemed, unlike any these reporters had heard from an oilman before. Their questioning blindsided Hubbert. Of course he'd give the talk, he told them.

After this encounter, Hubbert was furious. He welcomed critical thinking, but he could never stand it when people simply rejected new ideas without having thought them through. He got so worked up, he refused to go back to the hotel. Instead, he and Miriam skipped the conference's banquet and went out for dinner on their own, then to a movie. Finally somewhat relaxed, they returned to the hotel around midnight.

TWO DAYS LATER, on March 8, Hubbert sat on the stage in the Plaza Hotel's ballroom, where its tall windows gave him a panoramic view of the city, while he waited for San Antonio's mayor to finish his opening address.

When it was his turn, Hubbert would step up to the stout wood podium to deliver the conference's keynote speech, which he'd titled "Nuclear Energy and the Fossil Fuels." Before his turn came, though, someone signaled for him to get up and leave the room.

He rushed out of the hall to find he had an urgent call waiting from

Shell's headquarters in New York. At the other end of the line was an executive assistant in the public relations office, pleading with him to tone down his speech.

"That part about reaching the peak of oil production in ten or fifteen years, it's just utterly ridiculous," the executive assistant said.

Hubbert pushed back, saying his paper was simply "straightforward analysis." Finally he said, "Listen, the Mayor is giving a talk, I'm on next. Can we close this off?"

"Please tone it down some," the man said. Then he asked, "By the way, how many copies of that paper have been distributed?"

"Five hundred," Hubbert told him.

"Oh."

THE PETROLEUM ENGINEERS and other oilmen gathered in the ballroom that morning were in an amiable mood, chatting with friends and slapping each other's backs. When Hubbert suddenly rushed out of the room, he must have left many baffled, wondering if he'd return.

Soon enough he was back. As usual, he wore a plain gray suit and white shirt, but that day he'd taken extra care with his attire, with his bristly hair slicked back from his large forehead. He began his speech calmly and evenly.

His opening, however, was unusual. Instead of talking about oil, he spoke of Christopher Columbus. When Columbus first cast off from Spain in 1492, dragons still menaced the edges of the map. Beyond, entire continents awaited discovery and plunder. In the century after that famous voyage, Hubbert said, "several continents, a number of large islands, and numerous smaller islands were discovered." Even after all those finds, a seductive question still lured men to the sea: "How many more might there be?"

By the mid-twentieth century, though, there were no more continents to find. Explorers had sailed along most every shoreline and trekked across every continent, even Antarctica, and had planted flags on all but a few of the tiniest, most remote islands. In 1956 anyone who thought they could become another Columbus—a discoverer of vast new lands—was deranged.

"Likewise, for the petroleum industry the last century has been a period of bold adventure and discovery," Hubbert continued. "Whole petroleum provinces analogous to the continents have been discovered and partly explored." Explorers—from the major oil companies down to the scrappy independents known as wildcatters—had found a few hundred giant oil fields, the equivalent of large islands like Greenland or Australia. Smaller oil fields numbered in the thousands, akin to islands scattered across oceans. The industry owed its power to this unbroken record of successful discoveries. However, the age of oil exploration had to end eventually. Hubbert posed the natural question: "How far along have we come on our way to complete exploration?"

He then showed his sketch of a graceful bell-shaped curve, an idealization of the whole lifetime of America's oil production, past and future. At that time, the country was certainly still on the uphill part of the curve. Yearly production had been rising quickly and could still rise further. Nonetheless, based on Hubbert's favored estimate that the ultimate production would be about 150 billion barrels, this graph showed the production peak coming uncomfortably soon, around 1965.

The picture was much the same with natural gas. Based on a mainstream assessment of the ultimate production of America's natural gas, it too would reach a peak around 1970.

Anticipating that his audience would be looking for a loophole to escape these conclusions, Hubbert presented another scenario. Let's suppose, he said, the industry did manage to find a lot more oil—say, another 50 billion barrels, the equivalent of eight more East Texas oil fields, the largest yet found in the country. This would be a gargantuan task. But if they managed it and stuffed the additional oil into the bell-shaped curve, fattening it up, they could see how much more time that would give them before the nation reached the peak. Those additional 50 billion barrels would delay the peak only five years, until around 1970.

"It is difficult to escape the conclusion," Hubbert argued, "that the production of crude oil in the United States will reach its culmination

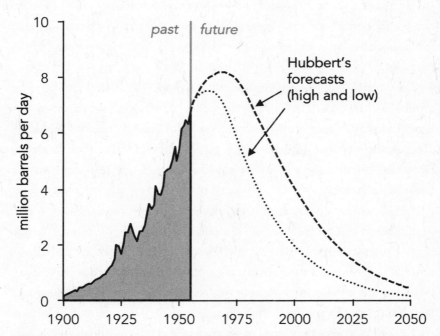

US oil production: Hubbert jolted the industry with his forecast that the nation's conventional oil production would peak between 1965 and 1970, then enter a long decline.

within a few years of 1965." Referring to his bell-shaped graph, "it is almost impossible to draw the production curve in any manner differing significantly from that," he said. The peak was near.

Hubbert also showed a similar outlook for world oil. Compared with his curve for the United States, when drawing a curve for the world, he had a lot more wiggle room. Nonetheless, assuming ultimate production of 1,250 billion barrels, it appeared the global peak would come in about a half-century, around the early 2000s.

These peaks would mark crucial turning points. "There is a vast difference between the running of an industry whose annual production can be counted on to increase on the average five to ten percent per year," Hubbert said, "and one whose output can be depended upon to decline at that rate. Yet this appears to be what the petroleum industry in the United States is facing."

If America's own production went over the hump and began to decline while consumption continued to soar, the country would become fettered with a large and growing dependence on oil from overseas, Hubbert argued.

But there was a more positive way of looking at the situation, he argued. When production began to decline, it wouldn't mean the petroleum industry would go out of business. Nor would America suddenly "become destitute of liquid fuels." Finding and extracting the remaining oil would become more complex but also increasingly important to the nation's well-being.

Also, besides crude, there were also other domestic sources of liquid fuels. They could cook down the oil shale in the Rocky Mountains, which could yield some 900 billion barrels, several times as much as the nation's crude oil. They could turn coal into gasoline. Large tar sands deposits in Canada, and smaller such deposits in the United States, could also add to the supply of liquid fuels. Such resources were large enough, Hubbert said, that "they should be able to supply all of our domestic requirements for at least a century or two."

These other sources could buy a lot of time. But in a truly long-term perspective, looking across several thousand years, a century or two was only a blip. Humanity would need energy sources beyond fossil fuels. Fortunately there was another possibility: nuclear power. The most promising approach, he argued, was a new type of reactor, the breeder reactor, which would not only consume radioactive ores like uranium but would also transform some of the uranium into a heavier element, plutonium, creating yet more nuclear fuel. Using these breeders, Hubbert said, was the only rational way to proceed with nuclear power, because otherwise they'd use up the uranium relatively quickly. If they did use breeders, though, America's uranium deposits would become an enormous energy source, several hundred times greater than all the nation's fossil fuels combined. The nation was on the threshold of a new era of abundant energy—but only if it made a momentous transition to nuclear power, on a scale unlike any effort ever undertaken.

In his long-term perspective, "the exploitation and exhaustion of the fossil fuels are seen to be but ephemeral events even during recorded human history," Hubbert concluded. "There is promise, however, provided that we can solve our social problems and not destroy ourselves with nuclear weapons, and provided that the world population can somehow be brought into control, that mankind may at last have found an energy supply adequate for its needs for at least a few centuries of the foreseeable future."

This last phrase was his dig at the American Petroleum Institute, which often talked of plentiful oil for "the foreseeable future"—but seemed to only ever look a decade or two ahead.

DESPITE HAVING BEEN mobbed by reporters before his talk, Hubbert's prediction got relatively little play in the media. Even in Texas—where discussing arcana of the oil business ranked second only to talking football—few newspapers covered Hubbert's warning. The *Corpus Christi Times* ran an article, "End of Oil Boom After 10 More Years Forecast," on page twelve, next to stories like "Walgreen to Open New Store" and "Four Girls Admit Kidnap Story a Hoax." Industry publications took Hubbert's prediction more seriously, however. The *Oil Daily* covered Hubbert's talk in a front-page article, saying it "was believed the first prediction of its type to be offered at an important industry meeting."

Actually there had been a similar prediction a couple of months earlier by Joseph Pogue, a veteran geologist with the Chase Manhattan Bank, a major financier for the oil industry. Similar to Hubbert, Pogue had forecast a peak of US oil production around 1970, but with an even sharper decline after the peak. However Pogue's prediction was based largely on the assumption—not explained or backed up—that the nation had already discovered half the oil it ever would.

Petroleum Week also gave Hubbert's forecast a fair and thorough treatment in a two-page article, "Is Oil Nearing a Production Crisis?" The article agreed that if his prediction were to come true, "the effect on every phase of the domestic oil industry will be profound." This article pointed out the similarities with Pogue's prediction and cited others who argued it was getting harder to find oil fields. Ira Cram of Continental Oil—one of those who, at the 1949 United Nations meeting, had presented an optimistic view—said that tapping America's oil fields was getting increasingly expensive.

There were many signs that the industry was having trouble maintaining its growth as usual. But Hubbert had stated his case more boldly than others, and it was based on an analysis much simpler and at the same time more rigorous. The only way to escape the conclusion that the peak was near, Hubbert argued in his talk, was if his estimate for the nation's ultimate production was "drastically in error."

WHEN HUBBERT RETURNED to the office on Monday following his talk, he discovered everything had come unglued. Shell was assembling a team to review his forecast. "It jolted the hell out of my own company," Hubbert recalled. "They were genuinely, honestly shocked."

They figured he had made a major mistake and wanted to see where he'd gone wrong. He was aware of the investigation and knew the production engineer put in charge of it, but they left Hubbert out of the discussions. "They just about had cat fits," he said.

In the meantime, Hubbert's invitation from Gershinowitz to lecture at the executive training program was put on hold. He usually enjoyed intellectual battles and often geared up for them. But in this case, his colleagues' aggressive reaction blindsided him. The oil estimates Hubbert had used were based on widely accepted work. "I didn't introduce any new figures," he observed. "I just showed them what the old figures meant."

PART 4

Houston, Texas, and Washington, DC, 1956–1973

Guesstimating

WHILE HUBBERT WAS WAITING TO hear the verdict of Shell's investigation into his oil forecasts, he received a letter from the veteran explorer Wallace Pratt. "I wish it had been possible for me to hear your address," Pratt wrote. "May I hope to receive a reprint in due time?" With pride, Hubbert mailed off a copy of his twenty-six-page report to Pratt.

Pratt quickly chewed through the paper and wrote a long reply. "The paper is meaty and thought-provoking," he said, admitting he envied it. "It has the flavor of a born and trained mathematician." Nonetheless, Pratt felt compelled to warn Hubbert. "I have long since come to be always a little nervous when one of my friends makes an estimate that the peak of oil production is imminent. You are too young a man to share the vivid impression I retain of earlier authorities who argued most persuasively as long as 50 years ago that we were even then approaching the end of our petroleum resources."

It was the same kind of argument Hubbert had heard from Arville Levorsen and others—except Pratt admitted he'd been one of those worried authorities himself. In the early 1920s, Pratt had become concerned that oil supplies could dwindle, so he used his position as president of the American Association of Petroleum Geologists to marshal its members to collaborate with the US Geological Survey. Together the two organizations assembled the best possible estimate of how much oil the country could ultimately extract. Their answer: 15 billion barrels. By that point, in 1922, the nation had already consumed over 6 billion barrels, so it appeared that within a couple of years it would have used up nearly half of the total. America's fast-growing consumption had already exceeded its production, so the United States began importing foreign oil. "This dependence," the study argued, "is sure to grow greater and greater as our own fields wane."

Within a few years, however, the industry made a huge number of discoveries enabled by transformative tools like the seismograph. It quickly became clear that these experts, from both industry and government, had been well off the mark. By the 1930s, the United States had a glut of oil, and instead of becoming a major oil importer, it became an exporter again.

Back in the 1920s, "I just couldn't see how it would ever be possible for us to produce more oil in this country than these well-informed authorities estimated," Pratt told Hubbert. "Yet, of course, we did." After riding the industry's ups and downs for decades, Pratt had changed. "I have come to realize that always in the past all of us have been too conservative."

Hubbert had used the latest, most rigorous estimates for the nation's ultimate oil production—but so had these earlier studies. "The best possible estimate of ultimate reserves isn't good enough," Pratt argued, "at least it never has been in the past."

ALTHOUGH HE DIDN'T mention it to Hubbert, Pratt had recently finished a report on the outlook for America's oil, commissioned by the US Congress. Drawing on his extensive contacts from a long career in the oil industry, Pratt surveyed nearly two dozen companies, asking each to create an estimate of how much oil the nation might ultimately yield. Based on the replies he received, he presented a best guess that the United States would ultimately extract 170 billion barrels of liquid hydrocarbons. (This catch-all category included not only crude oil and related condensates—what the industry usually meant by "oil"—but also so-called natural gas liquids, such as propane and butane.) Whittling down the number to only crude oil, it was the equivalent of about 150 billion barrels—the same as Hubbert's favored number.

This was a "conservative" estimate, Pratt wrote in his report. So he concluded that the nation had "abundant" petroleum and told Congress "the industry feels perfectly secure that it will be able to meet the demand it foresees over the next 20 years."

But even while issuing Hubbert a warning about the pitfalls of forecasting, Pratt was having doubts about his own outlook. Immediately after writing to Hubbert, Pratt had contacted Morgan Davis, a fellow geologist and longtime colleague. Davis was then vice president at Humble Oil, one of the largest US oil producers. In response to Pratt's survey, Humble had submitted a forecast that US oil production would continue rising for the next twenty years, at about the same rate as the previous twenty years. There were no limits in sight for supplies, Humble's report said. Instead, it argued, the main question hanging over the industry was whether there would be enough demand.

Given Humble's confident outlook, Pratt asked Davis how the company "has reacted to King Hubbert's prediction that 1965 will see us on the other side of the peak in our producing rate in this country." Pratt added, "I rate it as a much better paper than Joe Pogue's, even if it does reach much the same conclusion." Pogue, the geologist working for the Chase Manhattan Bank, had based his forecast on vague reasoning, with little evidence to support his case. Compared with Hubbert, Pratt told Davis, "all the rest of us were simply 'guesstimating.'"

But Davis wouldn't abide these pessimistic outlooks. As Pratt later put it, Davis "was perfectly furious and spurned King Hubbert's theoretical analysis of the data."

WHILE DEBATING WITH Pratt, Hubbert was waiting to hear the results of Shell's internal investigation into his forecast. In early April, only a week before the scheduled date for his lecture at Shell's executive training course, Hubbert finally got word of his company's decision. His lecture was on. Although his forecast was gloomy, Shell still wanted him to present it to its next generation of leaders.

Hubbert's boss never did divulge the results of Shell's investigation into Hubbert's methods. But whatever his colleagues concluded, it hadn't sunk him. "The whole thing had been a tempest in a teapot," Hubbert decided, "by people who didn't know what the hell they were talking about."

In April he flew to New York City, then drove north to Arden House, the sprawling ninety-six-room mansion Shell had reserved for the month-long training course. The building, the former home of a long-dead railroad tycoon, had imposing granite walls and a stone-lined interior featuring trophy heads of buffaloes, elk, and rams. To Hubbert, the place was "a marble mausoleum." Nonetheless, he thought it a good hideaway for two dozen budding executives, far from distractions.

For his afternoon session, he brought the same slides he'd shown in San Antonio. He projected his bell-shaped curves onto the screen and traced his pointer along them, forecasting a peak around 1965 or 1970. The trainees reacted, he felt, with "a certain amount of uneasiness." One man on the course thought Hubbert was the most pessimistic geologist he'd ever heard.

Nonetheless, after returning home, Hubbert received a letter from Shell's personnel director, who had organized the training session, thanking Hubbert for helping set the tone for the rest of the course. In the next round of training that fall, Shell wanted Hubbert to present his outlook again.

IN THE MID-1950S, many US oil companies were expanding into the Middle East, where prolific reservoirs yielded extremely cheap oil and production was rising fast. But Shell Oil was not part of the bonanza. As the US subsidiary of the European conglomerate Royal Dutch Shell, Shell Oil had to make its discoveries and its profits on US soil. Humble Oil was in much the same situation. Mostly owned by the industry's largest company, Standard Oil of New Jersey, Humble likewise operated only in the United States.

At the time, Humble and Shell vied for the title of America's largest oil producer. So these two companies, more so than others in the industry, had to contend directly with a growing surplus of oil on US markets. Oil production from the Middle East was flooding in, competing with domestic oil. Imports had risen by then to supply one of every five barrels consumed in the United States. Rather than worrying about whether US fields could supply enough oil, the main concern in the industry was whether Americans would burn enough.

The industry had a long-standing fear of gluts, which had repeatedly undercut their profits. There had been local gluts, like that in the Texas Panhandle region in 1926, when Hubbert first went to work in the oil fields. The Texas Railroad Commission had taken sporadic steps to try to quell such oversupply, with little success at first. But a turning point came with the 1930 discovery of the giant East Texas field, containing around 5 billion barrels, larger than any field yet discovered in the United States. The discovery had come as the nation was sinking deep into the Great Depression. Nonetheless, because the various companies were all draining the same fields, they drilled as fast as they could to extract as much oil as possible, before their competitors did. Faced with a flood of oil, the state—backed by a show of force from National Guardsmen and horse-riding Texas Rangers—finally made a successful push to regulate how quickly companies extracted oil. This allowed a slower, more careful rate of extraction, which Texas could adjust up or down depending on demand.

In this way, through a quirk of bureaucracy, the Railroad Commission wound up with almost dictatorial control over Texas oil fields. In the mid-1950s, Texas produced more oil by far than any other US state—and as much as the whole Middle East put together—so the commission's decisions affected the entire nation and the wider world.

By the mid-1950s, the commission regulated the production rate by taking advice from oil producers and refiners alike, then assigning quotas for each oil company, specifying that they would be allowed to produce only a certain percentage of what they theoretically could—a fraction known as "the allowable." When the total of these allowables was well below the producing capacity, the state wound up with "spare capacity," which could be called on in times of surprisingly fast growth of consumption, or in times of war.

Through the 1950s, US oil fields' production capacity had been rising faster than consumption. To adjust to this surplus, the Railroad Commission had been continually lowering the allowables. By 1956 the allowables were down to about 50 percent. In theory, Texas could simply open the spigots and flood the market.

WHEN MORGAN DAVIS of Humble Oil looked at US oil markets, he saw a glut. Hubbert's talk of limits made no sense to him.

Like Hubbert, Morgan Davis had grown up on a ranch in central Texas and become a geologist. But while Hubbert's relatives were poor farmers, Davis descended from plantation owners—the Old South's aristocracy. At six foot two, Davis was built like a linebacker and had stayed fit since his college days. In his midthirties, he had left behind geology fieldwork and remained in Houston, climbing the management ladder at Humble. When he took a training course at Harvard Business School, he had excelled—but was too relaxed, one of his teachers thought. Davis disagreed and felt his calm demeanor was one of his keys to success.

By 1956, Davis was in charge of Humble's oil exploration and was in the running to be the company's next president. A couple of months after Hubbert unveiled his prediction for US oil, Davis made a rebuttal at a Houston conference, disparaging Hubbert's work as being "in conflict with the views of most elements of the petroleum industry."

The venue was a meeting of the Society of Exploration Geophysicists, a gathering of the types of researchers behind advances such as reflection seismographs, which were responsible for much of the oil discovered over the previous few decades. "The whole scope of our activity," Davis told the audience, "has grown far beyond what even the most optimistic would have thought possible when many of us in this room first entered this business." There was reason for hope, Davis argued, because "much of this country has not yet been explored." Oilmen were then reaching out into new areas, sinking wells in the Gulf of Mexico more than ten miles from shore—so far out they couldn't be seen from land.

Given Texas's large spare capacity, there was no sign of an impending shortage, Davis contended. "It strikes me as strange indeed," he said, "that we should at this particular time have a rash of predictions that we should in so short a time as ten to fifteen years be unable to find and produce enough oil to meet demand."

Following this initial rebuttal, it seemed to Hubbert that every time he publicly presented his oil forecasts, soon afterward a rejoinder would appear from Davis or from his right hand, Richard Gonzalez. Considered one of the top economists in the oil business, Gonzalez was a staunch conservative. In a speech titled "What Makes America Great?" Gonzalez said, "Capital, energy, and freedom are the secret to progress at all times and anywhere." By "energy," he didn't mean verve or dedication. He meant fossil fuels.

In a 1956 article with a straight-shooting title, "We Are Not Running Out of Oil," Gonzalez claimed: "The dynamic petroleum industry has always confounded the prophets of gloom." To support that, he crafted a curious argument. He cast doubt on any attempt to predict the future of oil, saying that forecasts were not only difficult—they were impossible. "There is no way of proving today whether the peak of domestic ability to find and produce oil will be passed in the next ten to twenty years," he argued. Yet Gonzalez also reassured readers there was no need to worry about oil supplies. Following his repudiation of forecasts, Gonzalez presented his own outlook: "Domestic petroleum appears to have a reasonably bright future ahead."

DESPITE DOUBTS WITHIN Shell, warnings from Pratt, and critiques from the likes of Humble Oil, Hubbert's analysis did win accolades from others. Hubbert gave a copy of his talk to William Bradley, the

US Geological Survey's chief geologist, who quickly read it and asked for fifty more copies, so he could hand them out to the survey's geologists. "The thinking that you have shown in this address," he wrote to Hubbert, "is the kind of thinking that we would like to see more of in the Geological Survey."

Bradley then created two grade-eighteen positions—the highest non-managerial level in the Geological Survey—with one slot for Hubbert and the other for his friend William Rubey. Bradley sent one of his top geologists to offer Hubbert the job. But the pay was less than half what Hubbert earned at Shell, and he was most of the way toward securing a retirement package. He turned down the offer.

About the same time, in the fall of 1956, Hubbert received another letter from Pratt. The veteran oilman was known for being modest, even to the point of self-effacement. When presented with new evidence or ideas, Pratt could abandon his earlier stance and adopt a new view. In their earlier correspondence, the two men had reached a stalemate. By the fall, however, Pratt had shifted his attitude toward Hubbert's forecasts.

"I am a little embarrassed to tell you how good I think this paper is," Pratt wrote. "When I do so I sound too gushing." Pratt explained that he had been working on new reports—on the future of oil, and the impact nuclear power might have on petroleum—and found himself returning again and again to Hubbert's analysis. "Invariably, I have gotten more help from your paper than from any other," Pratt wrote. "I thank you accordingly. All this does not prevent me, of course, from hoping that you may be mistaken about the imminence of this country's peak production."

Toward the end of 1956, Hubbert also heard from Joseph Pogue, the lead author of the Chase Manhattan report that likewise foresaw a peak of US oil production within a decade or so. Pogue was curious how Hubbert's forecast might change if the country's ultimate oil production turned out to be significantly higher. Instead of 150 to 200 billion barrels, Pogue wondered, what if the ultimate production were 250 billion barrels? Would each additional 50 billion barrels delay the peak another five years?

That was roughly correct, Hubbert replied. But he expected that efforts to find more and more oil would suffer diminishing returns. To find another 50 billion barrels would require a gargantuan effort.

Hubbert saw a pattern emerging in the way his colleagues reacted to his forecast. Pratt clung to his hope for plenty of discoveries. Davis and Gonzalez implied that estimates would keep going up and up. Even Chase Manhattan, which had issued a forecast close to Hubbert's, seemed eager to find an escape hatch.

When writing his talk for the American Petroleum Institute meeting, Hubbert had thought there was a consensus about how much oil the United States could ultimately produce. But after his talk, that consensus evaporated. No one had made any huge discoveries or developed radical new technologies, yet his colleagues seemed to be fishing for higher numbers.

As Pratt had put it, by current methods, the "best possible estimate isn't good enough." Hubbert would have to come up with a new method.

IN DECEMBER, HUBBERT jotted down ideas in his lab notebook under the heading "Methods of predicting petroleum production other than based upon estimates of potential reserves." In addition to the production curve that he'd shown in his talk in San Antonio, he also drew a curve with the same shape but occurring earlier in time. "Discovery must precede production," he wrote. Obviously companies had to discover fields before they could extract oil from them. But Hubbert also hypothesized that the curve representing the history of discoveries would mirror that of the curve of production, both roughly bell-shaped.

The record of discoveries was based on the oil added to the industry's working stocks of oil fields, known as "proved reserves." Proved reserves were like a bank account. Discoveries of new fields were deposits to this account, and barrels extracted and sold were withdrawals from the account. As long as the deposits were larger than the withdrawals, the bank account's balance would increase. And over the long term, through the industry's entire history, that balance—the amount of proved reserves—had been increasing.

If discoveries reached a peak, then proved reserves would likewise peak somewhat later. In his notebook, Hubbert drew a third bell-shaped curve—his hypothesis for proved reserves over time. He figured proved reserves would reach a peak after the peak of discoveries and before the peak of production—about halfway in between.

In five pages, he had laid out an approach to forecasting that did not depend, from the start, on any estimate of the ultimate production. Instead it would draw on the industry's track record to date, looking at the long-term trends.

Each of these three curves had its quirks. Booms or busts in the economy could drive production up or down from year to year. Discoveries could be surprisingly large one year and disappointingly small the next, making the record of discoveries shudder like the needle of a seismograph picking up tremors. But the idiosyncrasies of each of the three

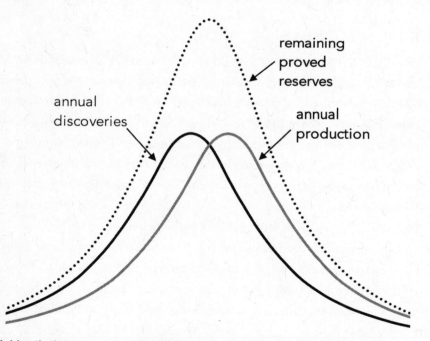

Hubbert's three-curve method: Hubbert sketched out a new approach to forecasting the future of oil, based on the record of oil discoveries, proved reserves, and production. This method avoided any assumption about the amount of oil that would be ultimately available.

curves should largely cancel out, Hubbert surmised. Together they would reveal long-term patterns more clearly than any one of them would alone. Importantly, both the discovery curve and the reserves curve would reach their peak before the production curve, providing some advance warning. After doing some calculations on the trends to date, Hubbert wrote in his notebook: "Production peak should follow peak of proved discoveries by about 12 years."

Just as Hubbert was working out this new method, he received a letter from Pogue at Chase Manhattan, suggesting he try looking at the history of discoveries as a way of predicting future production. Hubbert agreed it would be a good approach, since it would have "the advantage of not being based upon any assumption of ultimate reserves."

Hubbert hadn't worked out all the details yet, but he ventured a guess as to how the results would turn out. It would be difficult to know when the nation had actually reached its peak of discoveries, Hubbert explained, because the amount of oil discovered varied so much from year to year. The decline would only become clear in retrospect.

But Hubbert did see one sign that it was getting harder to find oil. The American Association of Petroleum Geologists collected all sorts of statistics on oil production and discoveries, and published lists of the oil fields discovered each year, grouped by size, from very small to very large. "While the number of fields found per year is being maintained," Hubbert pointed out to Pogue, "the number of larger fields is decreasing."

Then Hubbert made an intuitive leap. "This suggests to me that the crest of the rate of discoveries is probably about now," Hubbert wrote. He figured "discovery rates are due to decline."

The Beer-Can Experiment

WHILE READING *OIL AND GAS JOURNAL* in early 1957, Hubbert came across an intriguing article that he wanted to pass on to William Rubey. It described a spectacular oil well recently drilled in Iran. Two members of the drilling crew described how, on reaching a reservoir about 8,500 feet deep, "tremendous pressure beneath blew out mud and control equipment and spewed oil over the countryside at the rate of over 80,000 barrels per day for nearly 3 months," forming three lakes that together held more than a million barrels of oil.

Despite being swamped with work, Hubbert scrawled a four-page handwritten letter—his secretary was out with the flu—telling Rubey about this "very significant bit of information." The Iranian well proved that fluids could build up underground to enormous pressures, which came as welcome support for a hypothesis Hubbert and Rubey had been working on—one that could explain a century-old geological mystery.

Hubbert had gotten interested in this conundrum a couple of years earlier. While on his way from a conference in Italy to Shell's headquarters in the Netherlands, he'd taken a detour through the Swiss Alps, led by a young local geologist. There Hubbert got to see a famous site known as the Glarus overthrust, which showed a distinct horizontal line cutting through the mountainside, the boundary between two geological layers.

Oddly, the older layer was on top of the younger layer—the reverse of the pattern that prevailed in essentially all rock formations.

There were similar overthrust faults known in Scotland and Scandinavia and scattered across North America, from Montreal to Wyoming. In the mid-1800s, one prominent geologist thought he had a solution to the mystery, which involved rock folding over itself on a massive scale. This notion was so extreme, however, he feared if he published it, other geologists "would put me in an asylum."

Later researchers surmised that at some point after the younger rock formed, the older rock layer must have slid over it. But how could these layers—often a few miles thick—slide distances of ten or twenty miles over other rock layers? It seemed that there would be enormous friction between the two layers, so if pushed from one end, the upper rock layer would simply disintegrate before the whole slab started sliding. Or the upper rock slab could have slid downhill, pulled by gravity—but it seemed that slopes were never steep enough, over such a large area, to overcome the friction.

Hubbert had come to Europe to give talks on his work on fracking, explaining what happens when fluids were pumped into rock at high pressure. After looking at the Glarus overthrust, Hubbert realized that if water could naturally build up to high pressures underground, that would make the upper rock layers partially float. In turn, that would lower the friction between layers, allowing one rock layer to glide over another, like a hovercraft. This, Hubbert suspected, could be the solution to the long-standing mystery of overthrust faults. If water could build up to sufficient pressure, as Hubbert told a friend, then "out pops a ridiculously simple answer."

But could water attain high enough pressures? As yet, it wasn't clear.

AFTER RETURNING TO the United States in the fall of 1955, Hubbert had bounced the idea off his friend William Rubey, who'd worked for years on overthrust faults in Wyoming and Idaho. It turned out Rubey had independently arrived at the same explanation for these formations. Rubey had been gathering more evidence to support the idea but hadn't worked out the theory in detail. So he and Hubbert began collaborating on a paper.

Hubbert usually worked alone or with the help of an assistant. This was the first time in his career that he had written a paper with a colleague of equal standing. In the end, though, Rubey and Hubbert wound up dividing up the work into two related papers. Rubey took the lead on a paper describing the geological evidence, and Hubbert wrote another that spelled out the mathematical theory.

While writing these papers, they ran their hypothesis by some colleagues. One of their colleagues passed on a serendipitous observation that, in light of their theory, took on more significance. It involved the curious behavior of a beer can.

If you quickly emptied a can of cold beer, then turned it upside down, the air inside would initially be cool, due to the chilly walls of the can. But gradually the air trapped inside would warm up to the temperature outside. As the air warmed up, it also expanded—the same effect that pushes the pistons in a car engine, albeit occurring much more slowly. If this upside-down beer can was placed on a wet surface, then the increasing air pressure in the can would lift the can up—but the water would maintain

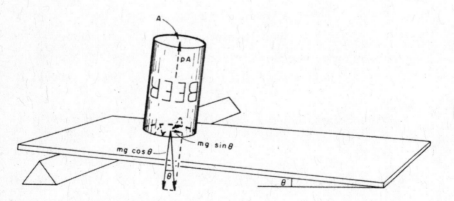

The beer-can experiment: A cold beer can sliding down an inclined piece of glass demonstrated the same forces behind Hubbert and Rubey's explanation of long-mysterious overthrust faults.

a seal around the edge of the can, keeping the air inside. If you then tilted the surface, the can—largely freed from friction with the surface—would glide downhill.

It was a party trick. But it was also an elegant display of the mechanism that Hubbert and Rubey had proposed for overthrust faulting. And as Hubbert had found with his work on fracking, a simple demonstration could go a long way toward convincing others.

Hubbert had tested out the beer-can experiment during a geology field trip in Nevada organized by his friend Thomas Nolan, who'd recently become director of the Geological Survey. "The idea is so fundamentally simple that there were no oppositional arguments," Hubbert reported to Rubey afterward. "The only question raised was whether the rocks could hold water at the required pressures."

That dramatic well blowout in Iran, along with other bits of evidence, suggested fluids really could build up to sufficient pressure. Hubbert and Rubey's hypothesis was not only physically possible, it was also realistic.

Presto! Chango!

RADAR AND THE ATOMIC BOMB, the mass production of synthetic rubber and penicillin—those were just a handful of the innovations that the Second World War had spawned. After the war came the polio vaccine, the hydrogen bomb (far more destructive than its predecessors made of uranium and plutonium), and the transistor, which promised to revolutionize electronics. In 1956 Congress passed legislation to build a nationwide highway network that, as one newspaper put it, would be "the greatest public works project ever undertaken by any country in recorded history." The United States had already launched a nuclear-powered submarine, and magazines like *Popular Science* and *Scientific American* envisioned atom-powered trains and cars—even airplanes. With such a string of technological and engineering marvels, it seemed no challenge was insuperable.

Yet there were still vulnerabilities. By the mid-1950s, most of Europe's oil came from the Middle East, by tankers traveling through the Suez Canal, a hundred-mile cut through Egyptian deserts that connected the Red Sea to the Mediterranean. In July 1956, Egypt's leader, Colonel Gamal Abdel Nasser, seized control of the canal—and in retaliation, Britain, France, and Israel attacked Egypt. The conflict wound up in a stalemate that kept the canal closed for months, into early 1957. In the midst of this, Saudi Arabia, one of the Middle East's biggest oil exporters, attempted to punish Britain and France by instituting an embargo, refusing to sell them oil. In response, the United States boosted its production, and oil companies rerouted tankers worldwide, managing to head off any significant shortages in Europe. Finally in the spring of 1957, the conflict was resolved and the Suez Canal reopened.

In the oil-consuming nations, many got the misleading impression that it had been easy to head off an oil crisis—that, as a Standard Oil executive put it, "all you had to do is push a button and everything was alright."

MEANWHILE MANY RESEARCHERS began chattering about vast amounts of oil remaining to be extracted. In early 1957 Resources for the Future, a DC-based nonprofit research organization, asked Hubbert to comment on a book they would soon publish, *The Future Supply of Oil and Natural Gas* by Bruce Netschert. The book came up with some "quite wild" results, Hubbert thought, such as that the United States could ultimately produce more than 500 billion barrels of oil. Netschert also forecast that US oil production could double by 1975, surpassing 16 million barrels a day, without any significant increase in costs. Hubbert wrote back a detailed critique of the book, saying Netschert "apparently has little firsthand information and experience with petroleum and natural gas," leaving him prone to believing some "fairly shaky speculations by poorly informed writers." Despite Hubbert's comments, Netschert stuck with his high forecasts and estimates, some of the highest on record.

Chase Manhattan also released a new report that drastically revised its view from just the year before. The bank raised its estimate for America's ultimate oil production from 165 billion to 250 billion barrels—enough to delay the peak about a decade, to around 1975.

In the autumn of 1957, *Oil and Gas Journal* surveyed several companies. Only one came out with a figure near Hubbert's favored number, 150 billion barrels. The rest ranged from 200 up to 500 billion barrels. The magazine concluded that the "generally accepted figure for US ultimate production was 300 to 350 billion barrels." Since the time Wallace

Pratt surveyed the industry less than two years earlier, the "accepted" figures had doubled.

Accordingly, the petroleum industry issued reassuring messages, often through the American Petroleum Institute. Hubbert had one of API's pamphlets, distributed to public schools to teach kids about oil, that said:

Q: Are we likely to run out of oil in the foreseeable future? Explain.

A: No. In spite of our large consumption, we are constantly adding to our known reserves. Modern engineering practices also increase the amount of oil obtainable from fields already discovered.

This was the API's "persistent propaganda line," as Hubbert put it. However, it was vague. It talked of the "foreseeable future," without saying how far into the future they could foresee. It belied an optimism that technology would always provide as much as people desired. It also confused the situation. The critical moment wasn't when the oil would "run out," but rather the much earlier time when production would peak, then begin to decline.

This kind of "misinformation" would hurt the nation, Hubbert told Pratt. "I have never been able to see how a forthright statement of the facts can do any possible harm to either the petroleum industry or the public," Hubbert wrote. The API was not "fooling anybody but the hopelessly ignorant and, perhaps, themselves."

Nonetheless, respected experts continued to issue optimistic assessments. Thomas Nolan, director of the Geological Survey, presented a similar refrain in a 1958 lecture organized by Resources for the Future, arguing new discoveries had "completely invalidated" the low oil estimates of a generation earlier. "It is true that we must face the eventual exhaustion of our oil fields," Nolan conceded, adding, "Hubbert has recently prepared an interesting and instructive discussion" of this issue. Nonetheless, Nolan argued, "we can view the possibility of exhaustion of even these reserves with some equanimity," since the United States could "produce synthetic liquid fuels from the tremendously large reserves of oil shales, tar sands, and in the still more distant future, low-grade coals." As the title of Nolan's talk put it, the nation had a bottomless well to draw upon: "The Inexhaustible Resource of Technology."

After Nolan gave this optimistic talk, Hubbert wrote him a long letter, arguing:

> The prediction of the future of petroleum production in the US is becoming a favorite industry pastime. The industry was happy and complacent over ultimate crude-oil reserves of the order of 150 billion barrels until I pointed out a couple of years ago that this implied a peak by about 1965.
>
> Subsequently there has been a considerable scramble to refute this conclusion, which can only be done by a drastic increase in the estimates of ultimate reserves. So, presto! chango!, on the basis of no significant new data, it is becoming fashionable to increase arbitrarily the figure on ultimate reserves by factors ranging from 1.5 to 2 or more.
>
> My reaction to this has been: Very well, let us not make any assumptions at all about ultimate reserves. Let us merely analyze the existing data contained in the curves of cumulative discoveries, cumulative production, and proved reserves.

DESPITE THIS INFLATION of others' estimates, Hubbert continued to forecast a fast-approaching peak of production of both oil and natural gas. He felt his newer three-curve method—drawing on the historical records of discoveries, proved reserves, and production—had bolstered his earlier outlook.

When the University of Texas in Austin celebrated its seventy-fifth anniversary, the university's president invited Hubbert to give a talk on the topic of "The Mineral Resources of Texas." If he hoped for an uplifting speech that celebrated the state's vast resources, he didn't know Hubbert.

In his talk in Austin, Hubbert spoke of peaks and declines in production of minerals such as salt and sulfur. Even groundwater could go through the same kind of peak and decline, since Texas residents were draining aquifers that had filled up over thousands of years. Much like fossil fuels, he said, extraction of this fossil water would likewise peak and then decline.

In forecasting the future of the state's oil, Hubbert showed his original method, which suggested Texas oil production would peak around 1965. Applying his newer three-curve method to Texas, he got an even more pessimistic result, suggesting the state's oil production would peak in just a few years, around 1962.

Facing the prospect of oil shortages, Hubbert again argued for a huge effort to ramp up nuclear power. But this time he also hailed the prospect of a whole other type of nuclear energy: the fusion of hydrogen atoms, the same process that powers the Sun. Fusion was "on the threshold of development," he claimed, and could supply vastly more energy than uranium-fueled power plants. Although he usually spoke of limits to resources, he said fusion held out hope of "unlimited supplies of energy," making it "potentially possible to meet all the State's needs for industrial energy as the fossil fuels decline."

Yet, Hubbert argued, even if Texans solved their energy problems, they would still face other hurdles such as overpopulation and limits to water. Addressing these issues would require more than technical solutions. During most of the state's history, he said, it had been "an essentially pioneer community with folkways appropriate to the frontier, and an implacable faith in the unlimited potentialities of exploitation and growth." So far that approach had worked. But it would have to change.

No major cultural adjustments have been required for the transition from the fighting of Indians to the wildcatting for oil, or from riding the open range of the Texas Panhandle to exploiting its ground water for its present intensive agricultural development. But very serious cultural adjustments may be necessary when there are no more frontiers to be occupied and no more virgin resources to be exploited.

Afterward, when Hubbert wrote to his old mentor Harlen Bretz about this talk, he said, "it was not exactly what the local Rotary Clubs or Chambers of Commerce would have preferred to have been told."

DURING THE SUEZ Canal crisis, US oil companies had boosted production to help Europe avoid shortages—and to cash in. But when the canal reopened, Europe once again had easy access to Middle East supplies. When the United States suffered a recession in 1957 and 1958, this further depressed demand for US oil, creating a glut.

In response, the Texas Railroad Commission had drastically tightened the taps. In the summer of 1957, the allowable had dropped below 50 percent, and by early 1958 it was below one-third—by far the lowest since the end of World War II. Texas wound up with huge spare capacity.

Meanwhile the US oil industry had faced other pressures. Labor costs were rising, and companies needed to drill ever deeper to find new oil fields, so overall costs were increasing. Even before the recession, petroleum industry profits had been withering. Humble Oil's profit margin slid from 22 percent in 1951 to 17 percent in 1956—and in 1958, squeezed further, the company laid off a tenth of its seventeen thousand employees.

In this situation, Morgan Davis—who'd recently been promoted, becoming Humble's president—focused on cutting costs and forecasting the demand for oil. Whenever the men from Humble issued rebuttals of Hubbert's forecasts, their arguments had been optimistic yet vague. They never issued a specific forecast for oil production or an estimate of the nation's ultimate production. "Throughout all this time," Hubbert complained, "Morgan Davis would never give a figure of his own. He'd always come up with great nebulous amounts of oil and gas, but never anything you could hang him with."

In November 1958, when Morgan Davis took the stage in Chicago to give the keynote speech at the American Petroleum Institute's annual conference, his outlook remained optimistic yet vague. But he did get specific in criticizing Hubbert's methods. Warning of "the hazards of prognostication," Davis recounted the recent rise in estimates for ultimate production—the jump from around 150 billion barrels to more than 500 billion barrels—as a reason to be skeptical about any forecasts. "Probably in no other field have so many intelligent prophets been confounded," he said.

Davis contended Hubbert's forecasts had gone wrong because he'd ignored a key factor: the nation's huge spare capacity. "The concern expressed as to a prospective shortage of oil has a hollow ring, particularly to those in Texas," Davis said. Instead of predicting a peak within five years or so, "Dr. Hubbert might have concluded that Texas production can continue to rise for 10 to 20 years."

Davis did concede that proved reserves "have failed to increase much in recent years." That might seem to support Hubbert's three-curve method, which suggested the nation's proved reserves would soon reach a peak—a harbinger of the peak of production. But Davis argued that proved reserves could be misleading. Looking at the estimates for how much oil could be ultimately extracted from each field, the numbers typically went up and up over time—so the estimates started out conservative, a "bare minimum," and over time the reserves grew, increasing by "a substantial margin."

"The public can be reasonably sure," Davis concluded, "that the United States will not soon experience a shortage of petroleum." When *The New York Times* covered his talk, the headline read "Bright Picture Painted for Oil."

BY LATE 1958, Hubbert had been extremely busy for months. Wrapping up his paper on overthrust faults, he'd grown exhausted from working on it seven days a week, while coping with persistent stomach upset. This marathon, he told Rubey, had been "remarkably like one feels in a bad dream when he is about to be overtaken by a tiger."

After submitting that paper to the *Bulletin of the Geological Society of America*, he took a couple of weeks vacation in Mexico over Christmas and New Year's. Returning to work, he was soon swamped with a steady stream of talks and other work. "If you never see me again," he told Rubey, "you will at least know I went down with all guns blazing!"

Hubbert had another confrontation in the works. He was to give a talk on predictions at the annual conference of the American Association of Petroleum Geologists, following talks by Morgan Davis and his colleague Richard Gonzalez. Although the invitation had come several months earlier, Hubbert started on the talk only a few weeks before the meeting. Nonetheless he decided to try to develop a new method for forecasting the future of oil.

Hubbert revisited something he'd discussed a few years earlier with Joseph Pogue of Chase Manhattan: the rate of discovery of major oil fields. *Oil and Gas Journal* published an annual tally of the discoveries of "large" fields—those estimated to hold at least 100 million barrels of oil. By the *Journal*'s count, in recent years the industry was finding fewer and fewer large fields.

Hubbert plotted out the cumulative number of fields discovered over time, creating a curve that went up and up. By that time there were 207 large fields on the *Journal*'s list. At some point, this curve would level off, when all the large fields were found. When Hubbert made his plot, it leveled off at the end, following a fairly smooth S-shaped curve, which pointed toward an ultimate count of about 220. That would mean that the discovery of large fields was almost complete.

"I couldn't really believe it," Hubbert recalled. Nonetheless he carried on with this method, estimating the ultimate amount of oil that all fields, large and small, might ultimately yield. The answer he got was 135 billion barrels—outside the range he'd used in his 1956 talk and significantly lower than the estimate he got from his three-curve method.

Hubbert—like many others—had gone wrong with such curve-fitting before. Back in the late 1930s, Hubbert had expected the US population would level off around below 140 million. But with the baby boom after the Second World War, America's population had soared, passing 150 million in 1950 and on track for 180 million by 1960.

His new result for the number of large oil fields seemed too pessimistic, which bothered him. "And yet that curve was dead on," he remembered, tightly fitting the history of discoveries. He put it in his talk.

THE WEEK BEFORE the AAPG conference where Hubbert would give his talk, President Eisenhower announced a major change in US oil policy. Eisenhower was concerned about growing dependence on imports and favored stockpiling oil for emergencies to boost national security. But when he'd floated this idea, it got no traction.

Powerful oil-backed politicians—such as two Texans, Sam Rayburn and Lyndon Johnson, who were the speakers of the US House and Senate—had been pushing for years for a different approach. They wanted quotas on oil imports to protect domestic producers from foreign competition, especially the Middle East's cheap oil. Arguments for import quotas often invoked national security as a rationale. But Secretary of State John Foster Dulles complained, "This business about the national security is a good deal of window dressing." The real rationale, he argued, was to "put more of the Texas wells into production and accelerate new drilling which will only happen if the price goes up."

Nonetheless Eisenhower felt he had little choice but to acquiesce to these oil interests. In March 1959 he placed a limit on foreign oil imports, capping them at 9 percent of the nation's consumption. His stated reason was "the certified requirements of our national security which make it necessary that we preserve to the greatest extent possible a vigorous, healthy petroleum industry in the United States," one capable of finding and developing new fields "to replace those being depleted."

This change of policy might not have much effect on how much oil the United States would yield over the decades to come, but it certainly would boost how much companies could profitably extract in the short run.

· · ·

"AT LEAST ONE controversy may arise," the Associated Press reported the day of the AAPG's annual conference. "Morgan Davis, president of Humble, is expected to declare there is plenty of oil to be found if there is a need for it," the article said. "Speaking later will be Dr. King Hubbert, who is known to hold opposite views."

Davis repeated his usual argument that technological development would enable US oil production to continue rising at 3 percent a year for the foreseeable future. Davis also critiqued Hubbert's new method of counting the number of large oil fields—before Hubbert had even made his presentation at the conference. Davis argued that many large fields weren't initially recognized as large, so recent statistics might under-estimate how many large fields had actually been discovered of late. Hubbert's method, however, had ignored such reserve growth. "There is sound reason," Davis said, to expect that the 1950s "will prove to have furnished large fields at about the same rate as the record performance of the prior twenty years."

Hubbert went ahead with his talk, showing his curve for the num-ber of large fields discovered so far and how it seemed to be leveling off. Technological improvements probably wouldn't drastically transform oil exploration, he argued. "I think we can rule out any radical improve-ment in geophysical techniques," he said. "Our seismographs are already so good that we are finding an ever-increasing number of smaller and smaller fields at greater and greater depths." The best hope for finding a lot more oil, he argued, was to improve the science of geology—that is, their ways of thinking.

At the end of Hubbert's talk, one man stood up to comment: Moses Knebel, Standard Oil's chief geologist, having taken over the job after Lewis Weeks retired. Knebel explained that about five years earlier, he'd completed an internal report for Standard that tackled the entire western hemisphere. Without divulging the details, Knebel said his results for the United States were similar to Hubbert's.

The next day Hubbert ran into Knebel between sessions at the con-ference. Knebel confided in Hubbert, revealing that Standard Oil's esti-mate for ultimate US oil production was 173 billion barrels—right in the middle of Hubbert's original range of 150 to 200 billion barrels. Knebel likewise looked askance at the estimates the industry had been coming up with in the past few years. It was "a rat race," Knebel told Hubbert, with everyone "trying to outdo one another in how large an estimate they can come up with."

A FEW WEEKS after the AAPG meeting, Hubbert wrote to Philip Abelson, the editor of *Science*, about his latest results. "There are powerful voices which are expressing strong disagreement with these conclusions," Hubbert said, "but they have yet to present any evidence that would justify my significantly altering them."

However, Hubbert soon realized he'd made a serious error that undermined his new approach. Morgan Davis hadn't presented much hard evidence—but his criticisms did hold some truth after all.

Going back over the data again, Hubbert tried plotting the data on the number of large fields discovered over time, comparing snapshots of the data from different years. Each snapshot of the data formed an S-shaped curve—but with each later snapshot, the curve kept inching upward. It "wouldn't stay still," as he later put it. "Every year when the new estimates were published, the fields had moved." Davis was right that fields could "grow," so some fields not initially thought to be large could later be recognized as large. This meant the ultimate number of large fields wouldn't be 220, as Hubbert had thought—it would be something significantly higher.

A few months after the AAPG meeting, Hubbert wrote to a friend, Preston Cloud of the Geological Survey, admitting that this oversight meant his earlier prediction had been "meaningless."

Hubbert hated being wrong. But when he felt he'd made a mistake, he wanted to be the first to admit it. He dove into fixing this problem. "I am now restudying the big fields," he told Cloud. It proved difficult to figure out when small fields grew into large fields, requiring "some digging around in obscure places" to find the right kind of data, since details on particular fields were generally kept secret. Although Davis and others talked about how fields could grow, it appeared no one had attempted to systematically analyze this effect, for all the nation's oil fields.

In the midst of this work, he remembered that his write-up for the conference was set to be published in the *Bulletin of the AAPG*. He couldn't let the error go into print. He phoned the journal's editor—as it turned out, at the last minute. "I pulled the paper," he recalled, "as the presses were about to roll."

The Rich Texans

DESPITE THE WIDESPREAD SENSE THAT the United States was facing an oil glut with no end in sight, one company had begun to pursue a new, radical method for getting more oil out of the ground. Richfield Oil, a small but aggressive independent oil company, was eyeing Canada's tar sands. These huge deposits—thick oil mixed with sand—had been known for centuries and had long been viewed as a back-up should conventional oil ever fall short.

The problem with the tar sands was that the oil had to be separated from the sand, and to do that, the tar had to be heated up enough to flow. This heating required huge amounts of energy. So in 1958 Richfield Oil proposed using nuclear bombs, detonated underground, to supply the heat. The company estimated that Canada's tar sands held "as much as 600 billion barrels of oil"—enough to satisfy current US consumption for two centuries. In talking with the Canadian ambassador, President Eisenhower enthused about the project.

The Atomic Energy Commission had recently begun funding similar types of endeavors, under the name Project Plowshare. One of the project's biggest supporters was the physicist Edward Teller—a veteran of the Manhattan Project, known as the father of the hydrogen bomb—who advocated many uses for nuclear blasts. Teller talked of sculpting a canal across Israel, a new link between the major Arab oil producers and

Europe, bypassing Egypt's problematic Suez Canal. He flew to Alaska to argue for blasting a new harbor there. In early 1959, in the Sunday magazine *This Week*, inserted in newspapers across the nation, Teller celebrated these varied possibilities in an article titled "Nuclear Miracles Will Make Us Rich." In that article, Teller also argued for using atomic energy to extract more fossil fuels. "There is enough oil locked in shale and tar sands to supply our current needs for hundreds of years," he claimed—and nuclear explosions would help get it out of the ground. The AEC didn't want to fork out money to support such oil projects, but Richfield was willing to pay for tests in the tar sands, so the effort moved forward under the name Project Cauldron.

Hubbert began looking into these proposals, which he considered "irresponsible and almost insane." His name for Project Plowshare was "Project Screw-ball." He read Teller's recent book, *Our Nuclear Future*, which gave a rough estimate that each nuclear bomb detonation might lead to two hundred cases of leukemia from the radioactive fallout. Teller dismissed this risk as uncertain and difficult to distinguish from other risks, such as exposure to the sun from living at high altitudes. But Hubbert found this attitude callous. Teller and the AEC, he felt, were "carrying out a determined propaganda campaign which has been based on a combination of suppression of vital information, and of depreciation of the danger of this fallout in direct contradiction to the best informed authorities on the subject."

One of the principal scientists involved in Project Plowshare—Gerald Johnson, the associate director of the Lawrence Radiation Laboratory, an AEC-funded site in California—came to visit Hubbert in June 1959 to talk about a proposal to blast oil shale deposits in Colorado's Green River formation. In an idea similar to Richfield's Project Cauldron, Johnson argued nuclear bombs could break up the shale and the heat would cook the shale, releasing oil that could be sucked out of the ground.

Hubbert told Johnson the technique probably wouldn't work, because the shale wouldn't fracture easily. There was a risk the nuclear blast could cause a blowout, spreading radiation across the landscape. On the grounds of both scientific feasibility and public safety, Hubbert told Johnson, he was "fundamentally opposed" to the whole Plowshare endeavor. Johnson was looking for industry partners who would pay for the shale blasts, but Hubbert informed him Shell would not participate.

THE OPTIMISM OF Teller and Johnson was widespread. During the 1960 presidential campaign, as the *Chicago Tribune* put it, the candidates had "the growth bug," all agreeing that when it came to economic growth, faster was better. The contest came down to Vice President Richard Nixon against Senator John F. Kennedy. Faster growth, they both argued, would fix all manner of problems, from raising poor Americans out of poverty to beating the Soviets in the arms race. The two opponents held different stances, though, on how to achieve higher growth. Nixon argued for free enterprise and lower taxes. Kennedy called for boosting government spending and promised to nearly double the growth rate, from 3 percent to 5 percent, saying, "a rising tide lifts all the boats" (a line one of his speechwriters borrowed from the slogan of a pro-growth lobbying group). Kennedy narrowly beat Nixon.

Amid this widespread fetish for growth, Hubbert's arguments about limits garnered little interest. Yet Hubbert did get the attention of one highly connected man, George McGhee, a geologist-turned-diplomat who had hammered out major deals in Iran and Saudi Arabia, helping mediate between the governments and US oil companies to maintain the American presence there. At the urging of a friend, Hubbert sent his "Mineral Resources of Texas" paper to McGhee, who liked it. McGhee then invited Hubbert to attend a meeting of the Philosophical Society of Texas, a century-old club for elites, which McGhee was president of. The group would hold its next annual meeting in December at remote Fort Clark, a decommissioned frontier base in southwest Texas, near the Mexico border.

Herman Brown, president of the massive construction firm Brown & Root, owned Fort Clark, which he used to entertain businessmen, influential newspaper reporters, and important politicians such as Senator

Lyndon Johnson, who'd recently become vice president–elect. Hubbert was optimistic about the opportunity of talking with these powerful Texans. Although they usually held staunchly to free market ideals, he felt that of late there had been an "intellectual ferment" that might make them receptive to notions of long-range planning.

Also, it turned out Morgan Davis was a member of the Philosophical Society and would be attending the meeting. Hubbert hoped to plumb Davis's mind, to understand the optimistic statements that he and his colleagues at Humble Oil regularly issued. "How much of this is put out as public relations fluff," Hubbert wondered, "and how much of it did they believe themselves?"

On the conference's second day, the afternoon session was devoted to a talk by E. B. Germany, an oilman, industrialist, and organizer for the Democratic Party—and McGhee asked Hubbert to make a commentary afterward. Germany—in his late sixties, white-haired, and fond of wearing Stetson cowboy hats—celebrated the "vast areas of good land" in Texas, the natural resources "almost without limit," the regions "rich beyond compare in oil, gas, sulfur." Echoing the state's long-standing independent streak, he called Texas a "self-contained economic entity" that was building a "self-sufficient standard of progress that may well become the showcase of America."

When Hubbert got his turn to speak, he compared Germany's speech to a doctor reassuring a sick patient and pumping him full of painkillers—while doing nothing to cure the disease. It was "replete with reassuring superlatives, but with very little usable factual information," Hubbert said.

Although Germany repeatedly referred to the enormous size of Texas and its resources, Hubbert said, "We should perhaps remind ourselves that whether a thing is large or small is a relative matter. In Alaska, I understand, the bartenders ask their customers whether they will have a 'regular-size' drink, indicating one about six inches deep, or 'Texas-size,' about three inches!" He laid out his forecasts that oil and gas production—in Texas as well as the United States as a whole—would soon peak. The state might appear to have enormous resources, but the nation also had an enormous appetite, so shortages could arrive surprisingly soon.

During the session, attendees recalled, King Hubbert and Morgan Davis had an "especially spirited" exchange. Afterward the Philosophical Society members and guests—almost exclusively men—were joined by their wives for dinner. Seated elbow to elbow on long benches, Hubbert wound up next to Vita Davis, Morgan's wife. Though Hubbert had known Morgan for several years, he'd never met Vita. King and Vita chatted amiably throughout the meal—until someone at the table called Hubbert by name. At that Mrs. Davis suddenly froze. "That so-and-so!" she cursed to herself.

Davis, it seemed, had been taking his skirmishes with Hubbert to heart—and venting to his wife about them. This amused Hubbert, who afterward wrote a thank-you note to McGhee, saying, "I thoroughly enjoyed the meeting."

Beforehand, Hubbert had developed high hopes that he'd be able to sway the state's elites. But as it turned out, his criticisms found no foothold. Afterward, he felt "pretty contemptuous of the whole thing," dismissing the attendees as "the rich Texans."

A Grand Survey

"WE NEED YOUR HELP," PRESIDENT John F. Kennedy told the scientists gathered at the National Academy of Sciences' annual conference. It was April 1961, three months into Kennedy's term in the White House. He had already launched an ambitious suite of efforts that won comparison with Franklin Roosevelt's prolific first hundred days in office, which had kicked off the New Deal. Among Kennedy's projects was a major new assessment of the nation's natural resources—a "grand survey," the press dubbed it, the most ambitious assignment the National Academy of Sciences had received in its century-long history.

At the academy's headquarters on Constitution Avenue, flanked by the State Department and the Federal Reserve Board, Kennedy explained in his speech that the government had to cope with "extremely sophisticated questions" that often "confound the experts." Nonetheless, he said, when policy makers faced "decisions which involve the security of our country, which involve the expenditures of hundreds of millions or billions of dollars, we must turn, in the last resort, to objective, disinterested scientists who bring a strong sense of public responsibility and public obligation."

Hubbert was among the scientists at the academy's conference. There he ran into an old friend, Edward Espenshade, who was by then chairman of the Earth Sciences Division of the National Research Council, an offshoot of the academy. It turned out Espenshade was privy to how Kennedy's "grand survey" was proceeding so far.

The survey had a major blind spot, Espenshade told Hubbert. Kennedy's request for the survey had said that "our entire society rests upon—and is dependent upon—our water, our land, our forests, and our minerals." Yet the panel of scientists assembled so far had no one to cover minerals—which, in geologists' terminology, included the nation's two major energy sources, coal and petroleum. Espenshade asked Hubbert if he'd like to get onto the committee to tackle minerals. Hubbert immediately said yes. Espenshade didn't have the authority to place Hubbert on the committee, but he could lean on the academy.

After that conference, Hubbert returned to Houston and composed a long letter to the academy's president, Detlev Bronk, a professor at the Rockefeller Institute for Medical Research. Hubbert explained he'd heard the natural resources survey wasn't planning to cover minerals, adding, "This impression may of course be entirely erroneous." But assuming it was true, he volunteered himself for the job.

"It is my thesis," Hubbert told Bronk, "that the world will exhaust the bulk of its high-grade mineral resources during the two-century period from 1850 to 2050." As a result, "very serious problems affecting the foundations of our economy are due to arise in the comparatively near future, and I see no way in which these can be circumvented except by national policies which are based upon a realistic appraisal of our mineral-resource situation." In a letter to Espenshade the same day, Hubbert went further, predicting "we are headed for something approaching a crisis in our mineral-based economy within the next few decades."

A couple of days later, Hubbert got a call from Bronk. Through one avenue or another, the message had gotten through, and Bronk invited Hubbert to join the committee. The academy had already lined up several top researchers, but Hubbert would be the committee's sole scientist with expertise on fossil fuels.

LESS THAN A week after Hubbert was invited to join the Committee on Natural Resources, he went to DC for the committee's first meeting.

There National Academy staffers laid out a general framework for the report. It was all centered on demand—that is, what a growing, increasingly affluent population would want to consume. Hubbert thought the approach was upside-down and that the starting point should instead be supplies—what would actually be available to consume.

Before Hubbert mustered an objection, another committee member did. Athelstan Spilhaus, an outspoken oceanographer, blurted out, "Horseshit!"

At this opening, Hubbert pounced. "Look, we're a committee that's supposed to advise the President on natural resources," he told the group. "Now, this committee is made up of very well-informed people. But not a one of us knows this whole field. So if we're going to advise the President, essentially our first job is self-education."

This mutiny caught Detlev Bronk by surprise, but he gave in. The committee members divided up the various resources among them, each taking responsibility for organizing a workshop on a particular topic. Hubbert, they agreed, would handle energy.

In setting up his workshop, Hubbert decided to cover fossil fuels himself. But for various other energy sources—from solar power to nuclear reactors to an exotic approach called magnetohydrodynamics—he drew up a list of experts he would invite to speak. He thought it might be hard to get these top researchers on relatively short notice, but to his surprise, everyone he called said yes. "It's very interesting," Hubbert mused. "To mention the President's name has a magic effect."

He still needed someone to cover nuclear fuels, principally uranium and thorium, so he asked Thomas Nolan, the Geological Survey's director, for a recommendation.

"Get Vince McKelvey," Nolan said.

Vincent McKelvey had been put in charge of the survey's assessments of uranium after the Second World War and had continued working on the topic since. Balding, round-faced, and fond of bow ties, McKelvey had a reputation as a careful geologist and a know-it-all. He also had aspirations of making a name as a big-picture, all-around intellectual figure. Having just been promoted to be an assistant chief geologist, McKelvey was clearly respected within the survey.

Hubbert took Nolan's word that McKelvey was the best man for the job, so he called McKelvey to invite him to speak at the workshop. But McKelvey said, flatly, no. McKelvey's reluctance was strange, Hubbert thought. After Hubbert reiterated that Nolan—McKelvey's boss—had recommended him for the job, McKelvey finally agreed.

In mid-July, when the Committee on Natural Resources met for Hubbert's workshop on energy resources, everything went smoothly—until the end of the second day, when McKelvey gave his talk. He was "like a bull in a china shop," Hubbert recalled. "He was plainly mad as hell about something." But Hubbert didn't know what.

Several months later, in January 1962, Hubbert returned to Washington for another meeting of the Natural Resources Committee. McKelvey tracked him down and handed him a thick slab of a report. It was, McKelvey explained, the Geological Survey's contribution to the academy's resource study. Although Hubbert had asked only for McKelvey's input on nuclear fuels, this report carried the comprehensive title "Domestic and World Resources of Fossil Fuels, Radioactive Minerals and Geothermal Energy." It covered every kind of energy that came out of the ground. When it came to assessing resources, it seemed that McKelvey felt he and his colleagues at the survey should be in charge—not the National Academy and the experts they'd assembled.

Curious to see what the Geological Survey's report had to say about oil, Hubbert flipped to that section, which turned out to be a two-page piece by Alfred Zapp. Helpfully, McKelvey had also included a longer report by Zapp, as yet unpublished.

Hubbert had been waiting a few years to see Zapp's work, which he'd heard about from his friend James Gilluly, Zapp's boss. According to Gilluly, Zapp had uncovered the reason for disagreements among various oil resource estimates. But how Zapp resolved things, and what his answer was, Hubbert hadn't been able to find out. Finally, in this report from the survey, Hubbert got to see Zapp's work. But on flipping through the reports, he was shocked.

Zapp pointed out that the amount of exploratory drilling done so far, about 1.1 billion feet, had uncovered a total of approximately 130 billion barrels. So on average, he calculated, each foot of exploratory drilling had discovered about 120 barrels of oil. Zapp then estimated it would require one well for every two square miles to achieve "near exhaustive exploration" of all the United States' promising oil areas. Assuming that future drilling would be about as successful as in the past, then with this amount of drilling, Zapp argued, there "seems little doubt" the industry could find enough oil to bring the total discoveries, past and future, up to 300 billion barrels.

Zapp's approach was extremely simplistic, Hubbert thought, as if oil exploration were "like plowing a cornfield." Zapp's calculations might be relevant if wildcatters drilled systematically, spacing their wells evenly from one side to the other of an oil-bearing basin. But no one searched for oil that way. There were much smarter ways to drill, using geophysical tools such as seismographs to pinpoint promising spots. Zapp's method seemed out of touch with the oil industry and oddly left little role for geologists or geophysicists to help. It seemed to assume that searching for oil was essentially random.

And yet, in that package of papers McKelvey gave to Hubbert, McKelvey had written a summary that said, "Those who are studying Zapp's method are much impressed, and we in the Geological Survey have much confidence in his results."

HUBBERT'S REPORT FOR the National Academy got delayed as he did other work—including a stint at Stanford, where he taught geology for the spring quarter, on loan from Shell. In April 1962 he finally returned to work on the academy report. On a visit to DC that month, he went to Nolan to complain about the estimates in McKelvey's report. "Look, that 300 billion barrels of oil is too much," Hubbert told him. "I can't accept that figure."

"Now, look," Nolan replied. "This came up from staff, so you go and talk to McKelvey."

When Hubbert visited McKelvey in his office, McKelvey said, "Hell, no—it wasn't 300." The figure was much higher. McKelvey pointed to a table of numbers in the report and said there were also another 290 billion barrels of "submarginal resources"—a category Hubbert had ignored as "trash." McKelvey explained that the total came to 590 billion barrels—one of the highest estimates on record.

Astonished, Hubbert tried arguing with McKelvey about it. McKelvey then called in another geologist, Richard Duncan, who brought a bunch of statistics and graphs. One out of five exploratory wells hit oil or gas—a ratio that had held steady for years, Duncan pointed out. So, he reasoned, it was safe to assume that future exploration would continue to be equally successful, discovering as many barrels for each foot of drilling as in the past.

"I didn't have any disproof," Hubbert recalled, "but I knew enough about this to be reasonably sure that it was seriously in error." Once back in Houston, Hubbert went over the AAPG's statistics, which suggested to him that the picture was not so rosy. Duncan had lumped in wells being drilled near known fields and so were much more likely to find oil or gas. When looking at just the "wildcat" wells—those drilled in unexplored areas—the success rate had dropped significantly since the end of the Second World War. Also, on average, the wells were becoming less and less productive. That meant it was taking more effort to find a new oil field prolific enough to be profitable to drill. Whereas in 1945 it took 26 wildcat wells to discover a new profitable oil field, by 1955 it required 47 wildcat wells.

After doing this research, Hubbert was sure the survey's researchers—McKelvey, Zapp, and Duncan—were mistaken. He wrote McKelvey a long rebuttal. Hubbert also reported back to Nolan, "Look, I can't even talk to this guy McKelvey. He's utterly irrational."

McKelvey's "mind was made up," Hubbert felt. "He knew the answers. So we just got nowhere at all. I mean, it was just a waste of time."

IN WRITING HIS report for the National Academy of Sciences, Hubbert ignored the work from the Geological Survey and made his own estimates and forecasts from scratch. He missed the deadline of June 1.

A couple of weeks later he was finally getting down to writing the report when he received a call from an oilman named Wallace Thompson, president of a Houston firm with a straightforward name, the General Crude Oil Company. Thompson wanted to talk about a major report in the works for the US Senate, called the National Fuels and Energy Study. Seeking feedback, the Senate had sent a draft of this report to many in the oil industry. Thompson thought the report—led by Samuel Lasky, a staff geologist at the Department of the Interior—seemed far too optimistic.

As Lasky's report put it, "Attempting to assess the Nation's oil resources on information that is publicly available is as frustrating as chewing on a mouthful of mashed potatoes: There is nothing to get one's teeth into." Lasky's report highlighted Zapp's work for the Geological Survey and also cited Netschert's 1958 book for Resources for the Future (which Hubbert had criticized for its optimistic outlook). Lasky's report also noted, "Other estimates place the volume ultimately recoverable as low as 150 billion barrels"—apparently a reference to Hubbert's work, although it didn't name him or cite his studies. Although the Lasky report mentioned these wide-ranging results, it said little about the studies' methods or whether one was more rigorous than another.

Lasky and his team were also frustrated in their attempts to estimate the oil industry's profitability. "We are told that the industry—contrary to other industry—has no idea how much money must be invested in order to find or prove a given volume of oil," the report noted. Although the report's authors included members of industry, as far as they could tell, "decisions are made on the basis of some sort of gambling instinct."

Despite such limited information, Lasky argued that the United States would ultimately yield over 400 billion barrels of oil and that the various estimates were "so large that we need not pursue the further speculation about the total amount in the ground." The report concluded "there is plenty of oil in the ground" and "the nation has the ability to be self-sufficient in oil if it so wills."

When Hubbert took a look at the Lasky report, he thought there were quite a few problems with it. He arranged to meet Thompson in a couple of days at the Ramada Inn to talk about it over lunch.

When Hubbert arrived for the lunch, he found they had company. Thompson had invited a few other industry men—including Richard Gonzalez from Humble Oil, who'd been hounding Hubbert for the past several years and who Hubbert considered "about the worst offender in the oil industry in propagandizing the public with overestimates." If Hubbert had known Gonzalez was coming, he never would have agreed to the meeting. But he did stay for lunch, listening while the other men bantered about the state of the nation, complaining about "that asshole in the White House" and how the country was going socialist. Hubbert resolved to keep his mouth shut.

When everyone finished eating, they finally got down to business. Hubbert explained that he'd gone over Lasky's report and the various studies it had drawn on and he felt the estimates were far too high.

Instead, the others talked about how to beat the coal industry. Coal supplied more than half of US electricity, whereas oil and natural gas together supplied about one quarter—and these oilmen wanted to grab a larger share of the market. Hubbert quickly decided that no one at the meeting, except perhaps Thompson, had any interest in arriving at accurate estimates. But he pressed them on it nonetheless.

"Well, gentlemen," he said, bringing things back to Lasky's report, "he comes out here with this figure, four to five hundred billion barrels, and you approve of that?" The whole reason the Senate had sent the draft around was to get the industry's endorsement of the estimates. "Suppose," Hubbert said, "those turn out to be seriously in error?"

The other men looked at each other. The study had been asked to make a projection out to 1980—and nothing could go seriously wrong in that time, they said. And if something did go wrong, one of them added, "I'll be retired by then."

Gonzalez visibly bristled. As Hubbert recalled, "I got one of the dirtiest looks I've ever seen. I think that guy could gladly have knifed me."

SOON AFTER THIS encounter, Hubbert finally finished his report for the National Academy of Sciences, drawing together all the different forecasting methods he'd developed so far. He reviewed the approach in his 1956 study, "Nuclear Energy and the Fossil Fuels," then pointed out how, following his prediction, there had been a huge inflation of estimates from industry and government alike, culminating in the Geological Survey's estimate of 590 billion barrels of oil. If there was no semblance of agreement on such estimates, Hubbert concluded, then they weren't trustworthy.

Hubbert's report then went over his three-curve method, drawing on the history of discoveries, proved reserves, and production, and avoiding any up-front estimate for how much would ultimately be extracted. When he'd first thought of the three-curve method in 1956, he'd figured the nation might, at that time, be at the peak of discoveries. Looking back, it appeared he'd been right. Annual discoveries had reached a peak in the mid-1950s and had been declining since. It seemed to him that proved reserves were similarly peaking right about then, in 1962. As for the third curve, he wrote, "the peak of production is expected to occur by about 1967 or earlier."

Judging from the rise and fall of these three curves, it looked like the contiguous United States would produce, in the long run, about 170 billion barrels—a figure that landed smack in the middle of the range of estimates he'd used in his 1956 forecast. (In his reports to date, he had excluded Alaska, while acknowledging it produced some oil and held potential to yield far more, as Wallace Pratt had estimated decades earlier.) Despite the recent crescendo of ever larger estimates, Hubbert felt the three-curve method bolstered his earlier forecasts, so he saw no reason to significantly change them.

He also presented another approach, which he'd first used in his 1959 "Techniques of Prediction" talk. But this time he did it right. In that earlier attempt, he'd looked at the number of "large" oil fields—those containing at least 100 million barrels of recoverable oil. As Morgan Davis at Humble Oil had pointed out a few years earlier, Hubbert had overlooked reserve growth.

In this new report for the academy, Hubbert highlighted this issue of reserve growth, explaining, "Every field which ultimately becomes a large field must go through an embryo, or incubation, stage as a small field before it ultimately hatches out as a large field." For this new analysis, he gathered data that showed how long it took, on average, to go through this incubation stage. Then he applied a correction factor to the fields discovered each year, estimating how many more large fields might have been already discovered but were not yet recognized as large. The correction made a huge difference to his estimate of the total number of large fields the United States would hold. When he'd first attempted this kind of forecast a few years earlier, he'd thought the ultimate number of large fields would be 220. But with his improved method, he arrived at a number more than twice as large: 460 large fields.

Along with some assumptions about how many more small fields might be discovered, he calculated a total amount of oil the country would ultimately yield: 175 billion barrels. That was satisfyingly close to the esti-

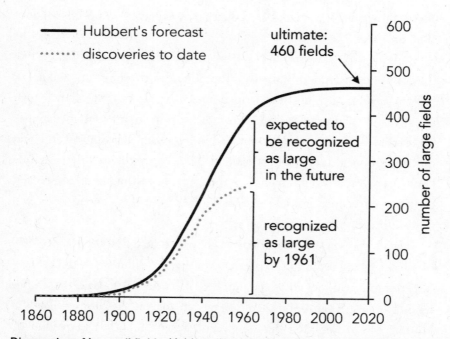

Discoveries of large oil fields: Hubbert developed another method for forecasting the future of US oil, drawing on the record of discoveries of large oil fields. However, many large fields were not initially recognized as large, so Hubbert devised a correction for this.

mate of 170 billion barrels from his three-curve method. With these independent approaches coming to about the same answer, Hubbert felt he was on the right track. The peak of US oil production would come, he felt sure, by the end of the 1960s. He included a rough "contingency allowance," arguing that perhaps the nation might yield another 50 billion barrels.

These forecasts could be thrown off somewhat, Hubbert pointed out. For oil, the main issue was how the nation would choose to use its spare capacity. As US oil production approached its peak, the Texas Railroad Commission might hold back production, trying to conserve what was left, which would mean production could fall short of his expectation and the peak would come earlier. Or Texas might open the taps, enabling production to continue rising and delaying the peak somewhat. Whatever the state's decision, Hubbert still expected the peak of US oil production to come by 1975 at the latest.

US natural gas production would likewise hit limits before long, Hubbert forecast. He showed how, over the past several years, estimates for US natural gas had gone through the same kind of inflation as the numbers for oil had. Using estimates he thought more reasonable, he forecast the nation's natural gas production would peak around 1975 or 1980.

Also, just as in his 1956 talk, he also showed a rough forecast for world oil production, based on the assumption that production would peak at double from the current rate, and that the world would ultimately yield about 1,250 billion barrels. If so, global oil production would peak around the year 2000.

AFTER HUBBERT SENT off his report to the National Academy, he flew to Denver to meet with Gilluly and Zapp. They debated their approaches all day. Though he'd known Hubbert for years, Gilluly was skeptical about Hubbert's forecasts. On seeing his graphs, Gilluly told him, "That's all doctored up with a lot of computer stuff."

"Nothing of the sort," Hubbert shot back. "It's all data, nothing is doctored."

They resolved nothing.

Hubbert then flew to Massachusetts, for another meeting of the Committee on Natural Resources—this time at a marine biology laboratory at Woods Hole, on Cape Cod. The purpose was to go over the reports by each committee member and finalize the whole package.

On arriving, Hubbert felt the mood was tense. As he recalled, one of the committee members, a biologist named Paul Weiss, sidled up to him. "You know, that report of yours," Weiss said, "I'm not sure that I could sign that."

"For God's sake," Hubbert snapped. "You're not supposed to sign it. It's my report."

Weiss, who was in charge of renewable resources, "didn't know one damned thing about oil and gas," Hubbert thought. With the way his colleagues were acting, Hubbert figured "something funny was going on before I got there."

On the last day of the meeting, another member of the committee, a chemical engineer from the Massachusetts Institute of Technology (MIT), also spoke out against Hubbert's conclusions. Arguing Hubbert's study was too pessimistic, the MIT professor said if it was published as it was, he would resign in protest.

"Gentlemen," Hubbert replied, "what do you propose to do with this report, burn it?" He said each member had been responsible for his own area of expertise, and if anyone could point out errors in his report, he'd gladly correct them. "Aside from that, the report stands. If you don't accept the report, I'll resign from the committee and publish it on the outside."

They begrudgingly accepted.

HUBBERT WAS FRUSTRATED with what he saw as knee-jerk rejection of his forecasts. When the Ford Foundation wrote to ask for his opinion of Resources for the Future—which the foundation had been funding, and which had published some of the most optimistic oil estimates on record—Hubbert sounded off. In a seven-page letter, he recapped the inflation of oil estimates over the previous several years, arguing:

Most oil-company sponsored groups as well as the managerial hier-archies of some of our largest oil companies are devoting intensive efforts at the present time to influence legislation in favor of the domestic petroleum industry. Consequently, they are deliberately putting out the highest estimates of petroleum reserves they can get away with. If they can find these estimates ready-made by dis-interested, non-profit research bodies, or by Government bureaus, so much the better. It is much more effective to use such figures than comparable ones of their own manufacture. And that is what is being done at the present.

Weeks later came the release of one such report: the optimistic report led by Samuel Lasky, which Hubbert had debated with Richard Gonza-lez and other oilmen over lunch several months earlier. The Lasky report received widespread media coverage, including in *The New York Times*. As Hubbert expected, the industry and its allies praised the Lasky report. The president of the Independent Natural Gas Association of America called it "an unbiased compendium of facts," and the executive vice pres-ident of the American Petroleum Institute deemed it "sensible, fair, and objective." The *Houston Post* said the study was "limited to facts and esti-mates, by and large, and shuns opinions and recommendations."

When Hubbert's report for the National Academy of Sciences was released in early 1963, the reaction was far different. Some of the media coverage took his forecasts seriously, including an *Oil and Gas Journal* article, "We've Found Half Our Oil and Gas, Hubbert Says." But given his critique of the many high estimates that had been issued by indus-try and endorsed by the Lasky report, he expected a fight with industry. Writing to *Science* editor Philip Abelson, he said, "I fully expect to have some of them after my scalp once they recover from the shock."

Soon the industry and its allies did go after Hubbert. The executive vice president of the Independent Petroleum Association of America told a Texas newspaper, "In the midst of problems growing out of surpluses, it seems strange indeed that we should now be asked to consider the prop-osition that the United States is in imminent danger of 'running out of oil.'" A *Houston Post* columnist argued, "If Hubbert's forecast were right, all the Kremlin would have to do would be to wait until 1967 and then let nature take its course as we fell by our own dependence on energy from outside sources."

Many of Hubbert's colleagues at Shell also remained skeptical about his forecasts. By the time his National Academy report was released, Hubbert was out at Stanford again, teaching for the winter. While there, Hubbert got word from Kenneth Deffeyes, a young Shell geologist he'd befriended. "None of the managers here seem to believe your predictions about the future of the oil business, even when the effects are already visible," Deffeyes wrote. He saw dire conditions ahead for research in the US oil business. Deffeyes had accepted a professorship in Minnesota, and several researchers at the Bellaire lab were likewise bailing out. "Some of us do not wish to sit around and watch a slow shrinkage," he added. "There are more exciting things to do."

As before, Hubbert's latest study brought a friendly note from Wallace Pratt. "Not the least commendable aspect of your estimates was the courage you manifested in publishing them in the face of a hostile prevailing opinion in the industry," Pratt wrote. "The resulting logic and analysis are so nearly invulnerable as to defy attack by ordinary mortals."

HUBBERT'S REPORT ALSO got the interest of Congress's Joint Committee on Atomic Energy, which was meant to serve as a watchdog for the nation's fast-growing nuclear power industry. When the committee invited him to testify, he flew across the country from California, arriving late the night before his appearance.

On starting his testimony, Hubbert was flustered, saying, "There was a little bit of a mix-up and misunderstanding about the prepared statement." It was Hubbert's first time testifying in Congress, and the National Academy had told him he wouldn't need to submit a statement, because his report would suffice. But then at the last minute they had changed their minds—perhaps when they remembered his report was 141 pages long—and asked him to write a summary. So, Hubbert told the congressional committee, "What I am giving you just now is a first draft written after 5 o'clock this morning."

The committee chairman, Senator John Pastore—a Rhode Island Democrat and former governor of the state—was known as a powder keg. Just five foot five, Pastore had a booming voice and was prone to fiery rhetoric. But he put Hubbert at ease. "I think you have a worthy case for an exception," Pastore said. "We are very happy to have you here and we are privileged indeed to get the benefit of your views. I would not let that disturb you."

Hubbert launched into describing various oil forecasts, pointing out "a rather wide range of disagreements among estimates that have been published during the last seven or eight years, so wide that one can only conclude that most of the estimates are rather seriously erroneous, no matter what the right answer is."

"What is the forecast in your report as to the life of our known petroleum reserves?" Pastore asked.

It was a common question. When geologists estimated how much oil remained, they often expressed it in terms of how long it would last at current rates of consumption. If Hubbert wanted to give that kind of answer, he could have said that US oil would run out in about fifty years. But putting it that way would run counter to his outlook that production would peak and then decline—and that this peak, the crucial turning point, would come surprisingly soon. Rather than answer Pastore's question directly, Hubbert began to describe his outlook, with an expectation of "an ultimate production of crude oil in the United States of somewhere in the neighborhood of 175 to perhaps 225 billion barrels. That is my figure. That is the figure that is favored in the report."

"At the rate we are using it, do you have any figure as to how long it will last?" Pastore asked again.

Hubbert replied simply, "Yes."

"How long?"

"I estimate that the United States will reach its peak of production about 1970, possibly a little before," Hubbert said.

"The peak of its production?" the senator asked.

"Yes, of the production rate," Hubbert replied, without explaining further.

"How long will our reserves last at that rate?" Pastore asked.

After some hemming and hawing, Hubbert said, "If I may say so, sir, that is not a question that is capable of a definite answer." He reiterated, however, that US oil was approaching a peak around 1970, and similarly world oil would peak around 2000.

The discussion then turned to nuclear power, the main topic of the hearing. "If we didn't have nuclear energy," Hubbert said, "I think we could seriously look forward to a complete decline of civilization within a few centuries just because we didn't have any energy to run it with. We would exhaust our high-grade metals and fuel supply and we would be back on sunshine and water power." He concluded, "Nuclear energy is coming just in time to save our lives."

"I would say your tone is one of alarm," Pastore replied.

"No, sir, I wouldn't say that at all," Hubbert said. "I don't think we are in an emergency situation and I don't think we need a crash program but neither can we afford to be complacent about the situation."

Going back over Hubbert's statements on oil, Pastore said, "You say we are apt to run out of this by the year 2000." (Pastore was making the common mistake of confusing the peak of production with the point of "running out.")

"No," Hubbert said. "What I said was that my estimate of the peak production of crude oil in the United States is about 1970." Hubbert failed to explain what he meant by that, however. Instead he jumped ahead to the implications, arguing, "We are not facing a crisis and I don't wish to create that impression." What he did want to get across was the long-term outlook, so that "the people who have to make such decisions will know about what speed we need to go in order that we don't get ourselves into very serious difficulties." Hubbert thought they needed to maintain an outlook that considered "a time scale of some hundreds of years in both directions," past and future. (That time scale, of course, didn't mesh with election cycles.)

Earlier in that day of testimonies, the undersecretary of the interior had told the committee that "the Nation's fossil-fuel reserves are ample to provide relatively low-cost energy in nearly all forms during this century and longer." Testifying alongside the undersecretary was Vincent McKelvey, who similarly summed up the outlook by saying, "known reserves minable, at or near present costs, will surely last until the turn of the century."

Senator Pastore also invited Morgan Davis of Humble to comment, and he sent in a statement agreeing with these Interior Department and Geological Survey estimates. "My analyses lead to the conclusion that there is no likelihood of a shortage of domestic oil for the next 20 years," Davis wrote, "if incentives for discovery and development are not impaired."

25

Penny Pinching

WHEN HUBBERT TESTIFIED AT SENATOR Pastore's hearing in February 1963, it was natural for him to drop in some of his long-standing complaints about nuclear waste disposal. As a member of the National Research Council's Committee on Waste Disposal, he'd spent several years working on the topic. Yet he and the committee had little to show for it.

For years, the AEC had been dragging its feet on dealing with nuclear waste, saying "they can't do this, that, or the other because they don't have enough money," Hubbert argued at the congressional hearing. "So my feeling is that this waste disposal thing needs to be given very considerable impetus."

BACK IN 1956, Hubbert and the other waste committee members had toured a couple of the AEC's sites, so they could see things firsthand.* At Oak Ridge National Laboratory, Hubbert had been shocked at what he

---

* Hubbert toured Oak Ridge National Lab in Tennessee and Brookhaven in New York. He had requested to be recused, citing the "responsibilities" of having high-level Q clearance. But in the end he apparently did get clearance or was allowed to go on the tour without it.

felt was lackadaisical handling of nuclear wastes. For liquid wastes with levels of radiation considered very low, the lab discharged them directly into rivers. For wastes considered low level, the lab had dug holes into shale deposits, creating ponds for the waste. The lab's notion was that the shale would filter out the more toxic and radioactive elements in the liquid, keeping them bound to the rock, and the remaining liquid could flow safely into the groundwater. But such shale could have many fractures, the waste committee had pointed out, so it could leak wastes before they'd been filtered. When questioned, the lab hadn't been able to account for where much of the liquid was actually going.

Waste that was still more radioactive—rated "intermediate level"—was being stored in concrete-lined steel tanks that were buried in the ground. Although the wastes were highly corrosive, the lab was "hoping that they wouldn't leak," Hubbert recalled. "We said to them, they damn well would leak."

When the Committee on Waste Disposal completed a report on what it had learned, in 1957, it had couched its observations and critiques in scientific terminology and diplomatic phrasing. It said its criticisms "should in no sense be regarded as criticism of officials responsible," since the facilities had been sited and run "during the exigencies of war." Yet, the report argued, the practices needed to change. "Unlike the disposal of any other type of waste, the hazard related to radioactive waste is so great that no element of doubt should be allowed to exist regarding safety," the committee concluded. "Stringent rules must be set up and a system of inspection and monitoring instituted. Safe disposal means that the waste shall not come in contact with any living thing."

For the most radioactive, "high-level" wastes, the committee argued, the most practical approach would be to dry out the liquid wastes, then store the remaining solids inside salt layers or salt domes, taking advantage of the special properties of those formations. They were "self-sealing": since the salt was able to slowly flow over years, if a crack opened in the neighboring rock, the salt would move into the gaps, keeping the nuclear waste isolated. Also, since the salt lacked fractures or other avenues for water to flow through, the domes' insides were bone dry. So wastes could be placed in metal casks, and in the absence of water, they should remain rust-free for hundreds or even thousands of years.

There were plenty of sites across the country where various disposal methods could be tested—and if promising, they could then be scaled up. However, the waste committee concluded that none of the AEC's major facilities were near the right kinds of geological formations for long-term nuclear waste storage. What was worse, the AEC's three largest plants, responsible for the bulk of the waste, were all located above massive fresh-water aquifers. But the AEC didn't want to have to transport the wastes away from these sites.

The AEC didn't seem to want to hear the waste committee's calls for much more stringent safety practices. As Hubbert recalled, the AEC "didn't like the criticism that we'd given them consistently right down the line." Following that 1957 report, the AEC had asked the committee to do some smaller bits of work—but never again asked for a major, comprehensive report. By 1963, the waste committee had sat completely idle for a couple of years.

IN HIS 1963 testimony to Congress's Joint Committee on Atomic Energy, Hubbert brought up the waste committee's report from several years earlier. "We set up our own criteria of safe disposal, which I see no reason to change," Hubbert said. "That is, this hot stuff should be isolated permanently from the biological environment so long as there is any danger from it whatever. If that takes a thousand years, so be it."

However, in the AEC's approach so far, Hubbert said, "I think budgetary considerations are involved, rather than taking a long-range view." The AEC didn't like the conclusion that its current sites were unfavorable for safely disposing of the wastes. Also, when choosing new sites for nuclear power plants or uranium refining facilities, the AEC wasn't factoring in waste disposal. To be disposed of safely, the wastes would likely have to be transported elsewhere—and that could be expensive, the committee calculated, since transporting them safely would require use of railroad cars with thick, heavy lead shielding. "In our first meetings with the AEC people," Hubbert explained, "this transportation problem was pointed out as being possibly the most expensive single item in a nuclear power program"—so expensive that it might make nuclear power unprofitable in some regions of the country.

The AEC seemed to be unconcerned about releasing some radioactive wastes into rivers and soils, as long as the levels of radiation were considered low, or the overall amounts small, Hubbert argued. But he countered, "We cannot afford to be careless merely because the quantities are small." If nuclear power were to scale up greatly, so would the volume of wastes. "It is not that it is dangerous this minute," he told Congress. "But when are we going to quit?"

IN MARCH, KING Hubbert left Stanford in his Technocracy-colored Mercedes—gray exterior, with black-and-red-leather interior, a custom order he'd picked up a few years earlier while on a trip to Germany. On his way back to Houston, he stopped in Denver to see Jim Gilluly, his friend who worked at the Geological Survey. On calling Jim, though, "he sounded kind of gruff and grumpy, and didn't sound as if everything was quite right," Hubbert recalled.

When King arrived the next morning, Jim's wife answered the door and broke bad news: Jim had suffered a heart attack two days earlier. "She wanted me to make this visit very quick," Hubbert remembered. "When a man's just had a heart attack, you don't talk about many things. You get the hell out."

The little time they did talk, they devoted to National Research Council matters. Gilluly was then chairman of the NRC's Earth Sciences Division. Hubbert had been slated to take over from him in the fall, several months off. But in the wake of his heart attack, Gilluly had decided to step down, so Hubbert would take over immediately.

Once back in Houston, Hubbert jumped straight into work for the Earth Sciences Division. His first priority was nuclear waste disposal. He'd grown frustrated with the AEC's "penny pinching," as he put it, with the lack of progress on this issue. He wanted to break the stalemate.

26

Returning Home

HUBBERT HAD ONCE TOLD A friend, half-jokingly, that to secure a retirement package from Shell, he'd have to endure many "years of servitude in the Houston climate." By 1963 he'd put in the required twenty years. The company's normal retirement age was sixty-five, which meant he could have stayed another five years. However, "I was perfectly happy to sign off at this stage," he recalled. "I'd done at Shell what I was capable of doing, except more of the same."

The Geological Survey had offered him a high-level position several years earlier, and ever since he'd had a standing offer. So he began talking with Thomas Nolan, the survey's director, to make the arrangements to take up a research position. "As you know," Nolan wrote, "we plan to give you a very broad assignment to formulate and carry out theoretical and experimental research." Hubbert would be considered a "lone worker"—the same position two of his friends, James Gilluly and William Rubey, had occupied before. Hubbert would be free to "follow the geologic paths that seem most promising," Nolan said—while expressing his hope that Hubbert would mentor younger researchers and work on structural geology and hydrology.

"I had no objection to that," Hubbert recalled, "but it wasn't what I was primarily interested in doing."

· · ·

AT SHELL'S BELLAIRE Research Center, Hubbert left behind a solid legacy. He'd been, as his colleague Kenneth Deffeyes put it, "the lead dog." Hubbert had helped design the lab and choose its staff. He'd made advancements that Shell had used in its search for oil, and he'd helped others solve vexing problems. On hearing the news of Hubbert's retirement, Wallace Pratt wrote to him, "What am I and the petroleum industry in general to do when we have questions that must be answered?"

Hubbert had run many courses within Shell, pushing his colleagues to think both rigorously and creatively about where to look for oil and how to get it out of the ground. When his colleagues attended these courses, he treated them like students and pushed them hard, which some found patronizing. But in these courses, he put in a lot of work himself. When the attendees turned in reports, he took the time to read them all and write extensive comments in response—unusual for someone in his high position.

Scientifically, Hubbert had many successes, as well as some failures. He pushed Shell to pursue his notion of hydrodynamic traps, but it hadn't panned out in the first region they looked, the Rocky Mountains. In the Permian Basin, straddling Texas and New Mexico, the ideas did lead to discoveries of hundreds of millions of barrels—but this was nowhere large enough to have a significant effect on Shell's prospects, let alone the course of US oil production. Besides helping get more oil out of the ground, Hubbert's theories on fracking allowed the company to control how it used waterflooding, a technique for recovering more oil from a field by pumping water into it. This had to be done carefully to avoid fracturing the rock, which could destroy the ability to sweep the oil out—and Hubbert's theory helped them understand what the reservoirs could withstand.

Hubbert's work on overthrust faults, done purely out of curiosity, also wound up having practical applications. Shell had run into trouble drilling into deep offshore oil fields that were below shale layers. The shale layers were "overpressured," just like the formations with high fluid pressure that enabled overthrusts. If Shell's drillers balanced those high pressures with heavy drilling mud, they often suffered lost circulation—

the same problem Hubbert had helped the company grapple with in the 1940s. Inspired by Hubbert's work on overpressured formations, Shell began looking for signs of those high pressures before it drilled into them. In the end, Shell was able to fine-tune its offshore drilling, leading to a drastic reduction in drilling costs and allowing it to tap fields that otherwise would have been too expensive to reach.

Hubbert set the tone in other areas as well. When word got around the Bellaire lab that he had donated money to civil rights activist Martin Luther King, Jr., it shocked his colleagues. At that time in Houston, "sending money to a black guy—nobody did that!" Deffeyes recalled. "When we heard Hubbert did it, several of us started sending money to Martin Luther King." Hubbert also badgered his colleagues—perhaps with less success—to delay having children, and to have at most two. (Hubbert and his wife had opted, of course, for none.)

Although Hubbert had made many contributions to the Bellaire lab, he'd always been a handful. Abrasive and hard-driving, he was difficult to manage or to work with. (There was that saying around the Bellaire lab: "Hubbert is a bastard, but at least he's *our* bastard.") Hubbert had found it difficult to keep an assistant for more than a year or two. The past five years, though, he'd had a brilliant assistant, Martha Shirley.* She was in an unusual position as one of the few female geologists in petroleum industry research. (When Shirley joined Shell, the sole other woman in Shell research flew over from where she was based, at a lab in Louisiana, to greet her.) Most companies wouldn't even hire a woman researcher, and when they did, they were treated as inferior, she recalled. Hubbert wasn't prejudiced against women, however. "He didn't care as long as I could do the math," she said.

But Shirley found Hubbert overly critical. When he received articles to review for the journals he edited, he had her check all the mathematics, line by line, to make sure the authors had done things correctly. "Nitpicking was his favorite thing," she recalled. He also criticized Shirley. "He

---

* On marrying later, Martha Shirley took the last name Broussard and went on to a long career in geology.

would pick on me because he didn't like the color of my dress," she said. "The nearest restaurant was a barbecue place, and he didn't like me going over there because then I would smell. There was just nothing he could be happy about." When he asked her to leave Shell and continue working for him, she declined.

Hubbert had other offers—from Stanford, the University of Houston, and Rice University. But he took the Geological Survey position in large part because he wanted to be back in Washington, DC, where he hoped to have "national influence."

Earlier, he had hoped that his long report for the National Academy of Sciences would make a mark. "I was trying to tell the President as plainly as I knew, and beyond him to tell the public that we had better start getting ready," Hubbert recalled. But then on November 22, 1963, after less than two years in the White House, Kennedy was assassinated. His successor, Lyndon Johnson, had his own ambitions for the nation—his Great Society program to boost civil rights and fight poverty.

In the end, Hubbert said of his work for the National Academy, "the influence of this report was as close to zero as possible."

"I FEEL AS if I were returning home," Hubbert wrote to Nolan near the end of 1963. Before joining, he hadn't asked for more clarity on exactly what the Geological Survey wanted him to do—or not do. "I should have but I didn't," he recalled. "I took it for granted that I would be continuing work of the kind that I'd been doing before"—studying energy resources.

The Department of the Interior put out a press release announcing that Hubbert would join the Geological Survey and would "virtually set his own investigative horizons as a research geologist and geophysicist, addressing himself to broad-scale problems." However, on starting there, Hubbert recalled, "I found out that there'd been a little politics behind the scenes to try to keep me out of oil and gas." Hubbert and the survey were in an awkward position. One of Hubbert's managers was also his staunchest opponent: the assistant chief geologist, Vincent McKelvey.

After joining the survey, Hubbert continued giving talks explaining the forecasts in his National Academy of Sciences report—and pointing out how they differed from those issued by the survey. The survey's chief geologist, William Pecora, finally gave Hubbert a warning, urging him to tailor what he said publicly. "It would be most inappropriate," Pecora wrote to him, "if your listeners, or the press, might misinterpret the thesis you might present as pointing up a conflict in official positions within the Department or the Geological Survey."

Hubbert had never tried to pass off his forecasts as "official." But as he wrote to a fellow petroleum geologist, "I am equally concerned for it to be clear that I am unable to accept or endorse the 'official' view of the Survey." Working for the government, he felt more hemmed in than at Shell.

Hubbert implored Nolan, as the survey's director, to sort out this situation. The controversy over oil estimates was "one of the more serious problems confronting the country today, and, willy-nilly, I am personally involved in it," Hubbert wrote to Nolan. "I am also distressed to see the Survey drifting (or being pushed) into a position which can hardly have any other result than to damage its reputation in the public esteem. I am deeply troubled about this and I am sure that you must be also, and I think the matter needs clarification."

The Zapp Hypothesis

AFTER HUBBERT HAD BEEN AT the survey for a couple of years, one of his colleagues, Thomas Hendricks, came to talk to him in his office. Hendricks wanted to clear up something about their former colleague Alfred Zapp, who had died a few years earlier. It was true, Hendricks said, that Zapp first had the idea of gauging success in finding oil by looking at the barrels discovered for each foot of exploratory drilling. And it was also true that Zapp had assumed that future exploration would continue to be as successful as in the past, an idea Hubbert thought of as "the Zapp hypothesis."

But according to Hendricks, Zapp had assumed this success rate would hold only for a while—not forever. The survey's estimate that the United States would, in the long run, yield 590 billion barrels of oil—the highest figure published in any credible report—shouldn't be credited to Zapp. McKelvey had always cited Zapp as the originator of the method, but it was actually McKelvey who'd assumed that the average success rate of the past would continue indefinitely into the future. After Zapp's death, McKelvey had continued using that assumption, generating ever-higher estimates.

At that time, Hendricks was wrapping up his own paper, based on Zapp's approach, but with an important tweak. Hendricks assumed that the success rate, measured in barrels discovered per foot of drilling, had been constant through the history of oil exploration—but that it would immediately begin to decline, eventually reaching zero. Hendricks's effort was only a slight variation on the approach by Zapp and McKelvey, and it generated a similarly high figure, that the country would ultimately yield 400 billion barrels of oil. And this, Hendricks argued, was actually a conservative estimate.

However, if the success rate of exploration had already dropped off from the historical average, then both McKelvey and Hendricks would be overestimating how much oil the country might find—not just in the long run but also in the short run. The problem was that the survey researchers were using overall averages rather than digging into the detailed historical record. They were, Hubbert thought, relying on numbers "obtained by pure assumption."

AT THAT POINT, King Hubbert had been "working at saturation" for two years. During that time, he'd wrapped up work at Shell—a period, he told his sister Nell, that was "hectic, if not a little nightmarish." King and Miriam had then moved away from Houston and undertaken a hunt for a new house in DC, which proved difficult. They'd shuttled back and forth across the country, spending each winter at Stanford, and Hubbert had completed his two-and-a-half-year stint as chairman of the Earth Sciences Division of the National Research Council. By 1966, he finally found time to put the Zapp hypothesis to the test.

"When I first read Zapp's thing, I never could understand exactly what the guy was doing," Hubbert recalled. "But in stewing over it for the next two or three years, it became finally crystal clear. What he'd done, if he'd followed through with evidence, was really a very superior method of analysis." Since Zapp had looked at the barrels of oil discovered, measured in terms of each foot of exploratory drilling, his method compensated for the ups and downs of the industry, when the price of oil fluctuated or the economy went through booms and busts. The barrels-per-foot approach, more than any other yet devised, would illustrate the ongoing race between advancing technologies and diminishing resources. If the barrels per foot were increasing, it would be a sign that technology was definitely winning the race. If that measure was decreasing, however, it could suggest that technology was not keeping up.

Rather than taking Zapp's approach of averaging over the whole history of the industry, Hubbert wanted to look at the record, year by year. The problem was that there was no ready source of data tying together the exploratory drilling from a particular year to the quantity of oil discovered that year. One major stumbling block was reserve growth—the problem Hubbert had run into, when trying to forecast the number of large oil fields remaining to be found. The total amount of oil recoverable from each oil field generally got revised upward over years or decades, and when the American Petroleum Institute reported these revisions, it lumped together fields young and old.

What Hubbert needed was data on discoveries that were backdated, so that any revisions to estimates for a particular field were credited back to the year that each field had been originally discovered. This information was generally kept private by companies. But Hubbert found a few studies that had access to that private data, and that listed the backdated discoveries for all fields discovered each year. The earliest was by the Petroleum Administration for War during World War II. He also found more recent work, published in 1961 and 1965, by the National Petroleum Council, an industry group set up to advise the federal government. Hubbert realized that these three studies, along with the initial estimates of new discoveries reported by the API, provided four snapshots of the fields discovered each year. By comparing one snapshot to the next, he could follow each cohort of fields—say, the set of fields discovered in 1932—and see how the estimates changed over the years.

It was like having the birth certificates for a bunch of kids, as well as snapshots of them as they grew up—from third grade to seventh grade to twelfth grade. Studying hundreds of kids, you could see how fast they grew, on average, at various ages—when toddlers, when young children, when teenagers. At first they grew quickly, and then gradually their growth slowed down, until they topped out at their adult heights.

Similarly, the various snapshots of oil fields showed how they grew

over time. The fields discovered in 1948, for example, were initially estimated to yield 800 million barrels over their lifetimes. In the next round
of estimates, made in 1959—when this set of fields was eleven years old—
the estimate had grown to 2.6 billion barrels, more than three times the
initial estimate. By 1963, after further revisions, they'd grown to a total
of 3.2 billion barrels—four times the original estimate.

When Hubbert plotted such changes on a graph, it revealed a regular pattern. Just like children, the fields grew quickly at first, and then
their growth slowed down. This pattern allowed Hubbert to calculate
how much each set of fields had grown over time—and more important,
to forecast how much more they would grow in the future. Over the long
run, from birth to old age, fields grew on average about sixfold. When the
fields "grew," nothing actually changed about them (other than getting
poked full of holes and gradually drained). What did change was the
industry's idea of how much oil the fields would yield—and Hubbert had
figured out a way to anticipate those revisions, decades in advance.

Once he calculated a value for the lifetime production from each
year's discoveries, he compared it with the amount of exploratory drilling
in the same year. That gave him a century-long historical record of the
amount of oil discovered per foot of exploratory drilling. This historical
record was far more detailed than what anyone at the Geological Survey
had been working with. It also factored in the reserve growth yet to come.
No one—neither at the survey nor in industry—had done this systematically, applied to all the nation's oil fields.

Hubbert found that in its early decades, the oil industry's success rate
had been above 150 barrels per foot of exploratory drilling. In the 1930s,
as the industry developed a more mature understanding of where oil
could get trapped and deployed radically improved exploratory tools—the
refraction seismograph, then the reflection seismograph—the success rate
increased, exceeding 250 barrels per foot. These technological leaps had
allowed companies to become more adept at identifying likely oil traps
before they put a drill in the ground. At the same time, however, companies had already discovered most of the shallow fields, so they had to drill

ever deeper to reach the remaining fields, racking up more exploratory footage. Even with ongoing advancements in seismographs, most exploratory wells still came up dry or encountered fields too small to develop profitably. In the late 1930s, by the time the industry had drilled some 3 million feet of exploratory wells, the success rate had started to decline. By the mid-1950s, the value had plummeted to less than 50 barrels per foot and remained similarly low through the 1960s.

Assuming this pattern of diminishing returns would continue into the future, Hubbert calculated the ultimate production for the contiguous United States (excluding Alaska). He did several iterations, tweaking this

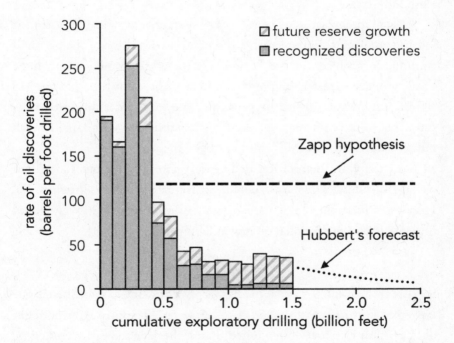

Success rate of oil exploration: The success of US oil exploration could be measured in terms of the barrels discovered for each foot of exploratory drilling. In addition to the recognized discoveries at any given time, there would also be future "reserve growth," with upward revisions to the fields discovered in years past. Hubbert devised a method for forecasting such revisions and showed that the success rate of drilling nonetheless had declined substantially over decades. His results were greatly at odds with Zapp's method, used by the US Geological Survey.

value and that value, to get a sense of the range of possibilities. His results kept coming out between 150 and 170 billion barrels—in line with his predictions by other methods.

Oil optimists—Morgan Davis, Vincent McKelvey, and many others—generally argued that new technologies would allow humanity to overcome limits to resources, a process with no end in sight. To counter this way of thinking, Hubbert pointed out in his new paper that "the present analysis is not based on assumptions of static technology." Instead his estimates reflected "the results of all of the improvements of exploration and production practices made during the entire history of the petroleum industry."

In extending past trends into the future, Hubbert argued, he assumed "that technological improvements comparable with those of the past will continue to be made in the future." He thought it would take a huge breakthrough—greater than any seen in the history of the industry—to significantly deviate from the pattern of the past. He allowed that his estimates were uncertain, but even so, "oil production as large as 200 billion barrels would be difficult to justify."

At that point the industry had already discovered about 135 billion barrels (including Hubbert's adjustment for future reserve growth). So, Hubbert argued, the US oil industry was certainly well on its way toward finding all the conventional oil it would ever find.

WHEN HUBBERT'S PAPER on discoveries per foot was published in the November 1967 issue of the *AAPG Bulletin*, as "Degree of Advancement of Petroleum Exploration in United States," he gave a copy to McKelvey. Here, finally, was hard data that Hubbert thought could resolve their long-running fight.

Hubbert was taken aback by the reaction he got. As Hubbert recalled, McKelvey gave a nasty growl: "I'll fix you!"

Up until this moment, Hubbert had approached the situation as a scientific debate—based on data, theories, and logic. Hubbert found

McKelvey's work on oil supplies to be simplistic, but still he thought it had been an honest attempt at forecasting. "In the early stages," Hubbert recalled, "I think there's no doubt that McKelvey was self-deluded—in other words, he believed his figures."

But with McKelvey's response to the "Degree of Advancement" paper, Hubbert's attitude flipped. "I could only interpret this as inexcusable dishonesty," he recalled. McKelvey was simply "trying to prove he was right the first time, and completely ignoring the evidence."

Hubbert was at a loss. "The logic was impeccable," he said about his own work. And yet, he added, "that logic was utterly ineffective. So all right, if rational analysis is ineffective, what can you do? Well, let's let the experiment decide, like Galileo and the falling bodies."

Galileo, according to an old story, had been debating other scholars over whether heavy objects fell faster than light ones. The answer seemed perfectly intuitive. Rocks plummeted to the ground while feathers floated down, so it seemed heavier objects had a tendency, even a desire, to fall faster. Aristotle, considered the greatest scholar of all time, had argued along similar lines.

But Galileo disagreed—and backed up his view with a clever argument. Everyone agreed that two identical objects, of the same size and the same material, would fall at the same rate. Suppose you tied the two objects together and then dropped them. Would they behave as a single, heavier object that would fall much faster than either one alone? If so, why would the addition of some string make this difference?

However, this logic didn't sway Galileo's contemporaries. To decide the matter, Galileo trudged up the stairs to the top of Pisa's leaning tower and dropped two balls, one heavy and one light. They hit the ground at the same time. Galileo was right.

Hubbert figured the world would soon provide such a demonstration. If he was right, US oil production would peak within a few years. If McKelvey was right, production could continue rising for another quarter-century. "Well, that's a pretty diagnostic criterion," Hubbert thought. "Let's wait and see what happens."

. . .

IN JUNE 1967, when Israel launched an attack on Egypt, the Suez Canal—Europe's connection to Persian Gulf oil—was once again shut. Compared with the earlier closure of the canal in 1956, this time more was at stake.

In the intervening years, Europe's consumption of Middle East oil had tripled. Also, the major oil exporting nations—including Saudi Arabia, Kuwait, Venezuela, and others—had banded together to form a cartel, the Organization of Petroleum Exporting Countries (OPEC). Its founders had been directly inspired by the Texas Railroad Commission's success in keeping prices up to support US oil production. OPEC wanted to do the same on a global scale, with the aim of usurping the role that had long been held by the major oil companies, most of them American. The Arab members of OPEC also aimed to control their oil exports to achieve political ends—in particular, to try to push the United States and Europe to temper their support of Israel.

With the outbreak of war, many Arab members of OPEC cut their production drastically, a sudden drop of production of 6 million barrels per day—the equivalent of about two-thirds of US production. These Arab nations also placed an embargo against the United States and the United Kingdom. In response, the United States ramped up production, and some OPEC members, including Venezuela and Iran, broke ranks and also ramped up production. After a couple of months, the Arab nations gave in, raising production and lifting the embargo. It was a test of the new group's power, an attempt to wield oil as a weapon. It ended a complete failure.

IN MID-1968, AS US elections approached, Hubbert got word that the State Department was eager to see his latest forecasts, to use in a report for the next US president.

President Lyndon Johnson had pushed the United States further into war in Vietnam, but his policies grew increasingly unpopular, and he decided not to run for reelection. Robert Kennedy then entered the race, quickly becoming the Democrats' front-runner, winning some early primaries—but then, like his brother, he was assassinated. The leading Democratic candidate wound up being Hubert Humphrey, the lackluster vice president. With antiwar protests and civil rights demonstrations sometimes turning violent, and the assassination of Martin Luther King, it seemed to many the country was growing lawless. Promising to restore order, Richard Nixon entered the race, contending once again for the Republicans' nomination.

Hubbert was then in the midst of working on a new report for the National Academy of Sciences, an update of his 1962 report, but didn't have it ready. He passed along word to the State Department that the new report would have "no significant changes in either the data or conclusions." He still expected US oil production to peak very soon.

Before the 1967 Middle East conflict, Texas had been gradually raising its allowables to feed America's fast-growing consumption. After that conflict, the state kept its allowables above 40 percent—the first time in a decade they'd remained that high. The United States was having difficulty boosting production enough to keep up with rising consumption and to hold imports at tolerable levels.

Around this time, one of the US government's top oil experts, James Akins, grew deeply concerned. Akins was head of the US State Department's Office of Fuels and Energy and also an expert on the Middle East. He helped coordinate with the Organization of Economic Co-operation and Development (OECD), a coalition of the non-Communist world's industrialized nations, which made them also the world's largest oil importers. The OECD had an Oil Committee that was meant to help its members coordinate to avoid shortages. During the 1967 Middle East conflict, the United States had called for the Oil Committee to declare an emergency, but European members voted against this, arguing that there was little reason to worry. When the Arab cutback and embargo fizzled, the Europeans seemed justified.

In late 1968, Akins traveled to Paris to attend another Oil Committee meeting, where he made a surprising admission. He warned them that America's spare capacity was almost gone.

This "was a triggering point," recalled Ulf Lantzke, then head of West Germany's Energy Department. "From that point onwards, I was trying to turn around energy policies in Germany." But Lantzke found it difficult, "so deep-seated was the political belief that energy supplies were no problem."

Unlike the Europeans, the US government did show some signs of concern. Richard Nixon won the 1968 presidential election, and once in office he launched a major effort called the Cabinet Task Force on Oil Import Control. This assessment, overseen by Nixon's top advisers, would take a fresh look at America's fast-rising dependence on foreign oil.

28

Ruthless

BY THE LATE 1960s, the United States seemed to be heading for "an inevitable age of nuclear power," as *The Washington Post* put it at the time. But the Atomic Energy Commission faced increasing objections to its secrecy and its processes for approving new nuclear power plants. These challenges only grew stronger after a fire broke out in May 1969 at Rocky Flats, an AEC site outside Denver that was the nation's primary site for processing plutonium for nuclear warheads.

For a few months, the AEC kept news of the fire under wraps. But after it briefed a Senate subcommittee behind closed doors, word leaked. *The New York Times* reported on the fire, then other outlets piled on, digging deeper into the story. Small fires, it turned out, were a regular occurrence at Rocky Flats, since plutonium was prone to bursting into flame. The large blaze had broken out over a weekend, in a building tended by a lone watchman. It caused tens of millions of dollars in damage and burned up enough plutonium to build dozens of bombs. When plutonium burns, it turns into plutonium oxide—chemically changed but still equally radioactive. The fire left behind some 330,000 cubic feet of contaminated waste, enough to fill sixteen Olympic-size swimming pools.

The AEC had been sending such waste to its site in Idaho, the Nuclear Reactor Testing Station, for years. But given the agency's secrecy, this was news to many. On learning of this, Frank Church, a Democrat representing Idaho in the US Senate, demanded more information.

Much of the radioactive waste sent to the Idaho site was simply buried in pits. The AEC told Church that this was safe, arguing that the region got so little rainfall that the waste would stay put, without reaching water supplies. Skeptical, Church called on the Geological Survey to study the situation.

When members of the National Research Council's nuclear waste committee learned of Church's interest in this issue, they saw an opportunity to get attention to a report of theirs that had long languished.

SEVERAL YEARS EARLIER Hubbert had used his position as chairman of the National Research Council's Earth Sciences Division to try to pressure the AEC to give the waste committee more work to do. The committee could do studies only at the AEC's bidding—but it seemed the AEC wanted the committee around only for show. After its initial long report in 1957 and a few other bits of smaller work, the committee had become essentially defunct, existing in name only.

Finally, in 1965, Hubbert gave the AEC an ultimatum. He told one of the agency's representatives, "Either there's something for the committee to do, or discharge the committee." The AEC agreed for the committee to do one more study. For the first time in the committee's decade-long history, it toured all of the AEC's major waste disposal sites. The committee members stopped in Oak Ridge, Tennessee, where Hubbert had earlier seen radioactive liquids fed into seepage ponds, which were meant to filter out the radioactive and toxic elements. In Kansas, they visited salt mines, the site of waste storage experiments dubbed Project Salt Vault. Last came Hanford in Washington state, the most contaminated AEC site.

After the tour, the waste committee wrote another report. It followed the same guiding principles as in its earlier work, the foremost being, "Safety is a primary concern, taking precedence over cost." While noting there had been some development of promising techniques for disposing of waste, the committee maintained its recommendations about the best overall approaches, such as putting the waste in salt deposits, and reiterated its criticism of the AEC's practices to date.

In 1966, when the committee turned in its draft report, the AEC reacted more strongly than before, saying the committee had exceeded its remit, which was to comment only on possible research into nuclear waste disposal rather than current practices. So the AEC said all references to current practices should be deleted from the report.

Such cuts would eviscerate the report, the committee felt. It refused.

The waste committee then received a rebuttal from all the way at the top of the AEC: Chairman Glenn Seaborg, a Nobel Prize-winning chemist. Seaborg said the AEC had long been aware of the importance of safe handling of radioactive waste. He cited the assessment of the Congressional Joint Committee on Atomic Energy, which concluded the nation's nuclear waste practices "have not resulted in any harmful effects on the public, its environment, or its resources." Seaborg also sent an internal AEC report stating that the cost of safe handling of radioactive wastes would be low—less than one percent of the total cost of nuclear power.

However, what the AEC considered safe was different from the waste committee's standard. And though Congress's Joint Committee had given the AEC its stamp of approval, Congress wasn't actually providing any oversight, Hubbert thought. In reality Congress was "entirely collusive," he said. "There has never been an agency any more ruthless than the AEC. They were a law unto themselves."

SO IN 1969, when the waste committee heard of Senator Church's interest in nuclear waste disposal, it told Church about its report, and about how the AEC had steadfastly refused to release it. The committee didn't divulge what the report said. The fact that it had been suppressed was enough of a red flag.

Church demanded the AEC explain the situation. What he got back was a condescending letter from Seaborg, saying the AEC commissioned many reports that it didn't release because they concerned internal operations. Also, Seaborg argued, the waste committee had gone "beyond the requested appraisal." Rather than simply recommending topics for further study, it had looked at disposal practices in use, "concerning which the Committee had been given only a short briefing." Seaborg seemed to be arguing that the AEC hadn't released the report because it wouldn't be of interest to anyone else, and that in any case the committee didn't know what it was talking about. He didn't state anything specific that the committee had been wrong about, saying only that AEC staff had "many criticisms" of the committee's report. Seaborg still refused to release it.

UNDER INCREASING SCRUTINY, in September 1969 the AEC made one of its first attempts to be more transparent, having Seaborg and other commissioners and staff deliver speeches in a day-long event in Burlington, Vermont. The talks wound up being long-winded and overly technical. In a panel session, the AEC faced critics who lambasted the commission's safety record. The AEC didn't come off well. One commissioner thought the whole thing an "utter disaster."

Meanwhile, an uprising was brewing in Minnesota. The AEC was set to approve construction of a nuclear power plant upstream from the metropolis of Minneapolis–St. Paul. But the state of Minnesota issued its own rules for the release of radioactivity that were stricter than the AEC's—and which the AEC did not want to abide by. Since the AEC was meant to have jurisdiction over nuclear power in the entire nation, this showdown was the first to directly challenge its authority—and to question its double-headed role as both promoter and regulator of atomic energy.

As the confrontation continued, the University of Minnesota organized a conference to discuss the risks and rewards of nuclear energy, which would give AEC representatives another chance to present their case to environmentalists and others demanding stricter safety laws. One of the organizers invited Hubbert to give a lecture that would put nuclear power in a larger perspective. As a strong supporter of nuclear power but harshly critical of the AEC, Hubbert straddled the two camps.

Still frustrated that the AEC had buried the waste committee's report, Hubbert packed his copy of it. In Minnesota, he planned to blow the whistle.

He took the train from DC, arriving in Minneapolis the day before the conference. He'd planned to take a taxi to his hotel. Instead, "I found myself surrounded by a bunch of AEC people and a private limousine," he recalled. "And then the next morning, the same thing," with a limo to take him to the conference. He thought this was strange—until they arrived at the university for the conference, where he saw protesters carrying placards. He surmised the AEC was isolating the conference speakers, to prevent "anybody talking to us, or us talking to anybody."

Out of all the conference's speakers, Hubbert, then sixty-six, was the white-haired "respected world authority on man's resources," as *The Washington Post* described him. In his talk, he called fossil fuels a "jackpot of energy" that wouldn't last. While he strongly supported nuclear power, even that new energy source wouldn't allow infinite growth. "It has become mandatory to stabilize the world's population and industrial activity at a level that earth resources can stand," he argued. "We may even have to drop the population back to some livable level."

Later, when the conference opened up for discussion, Hubbert spoke up first, saying, "According to AEC sources, there is a promise of an acute shortage of uranium-235 in about 25 years," due to consumption by the nuclear reactors then being built, he said. He'd always considered it essential to ramp up breeder reactors as quickly as possible, to avoid using up the nation's uranium resources. "Yet, the breeder program has been extremely slow getting started," Hubbert said. The nation, he charged, was following a "very short-sighted policy."

"Dr. Hubbert is entitled to his views," replied James Ramey, one of the AEC's commissioners. "He has been trained in geology and knows a great deal about natural resources, but he is not an expert in nuclear power."

Hubbert said nothing in his defense.

Throughout the meeting, Hubbert had many opportunities to bring up the waste committee's report. A law professor from George Washington University spoke of how "the atomic energy establishment is prone to dismiss those who are concerned about the health and safety implications," treating them as "ignorant of the facts." That could have been a prompt for Hubbert to mention the suppressed report. But throughout the whole conference, he never did.

Since the conference's talks and discussions would be published as a book, after the meeting Hubbert sent the organizers a seven-page "addendum" to his comments. Although he'd failed to blow the whistle during the meeting, in this commentary he told of how publication of the waste committee's report had been "persistently withheld by the Atomic Energy Commission because of the report's criticism."

When the editor of the conference proceedings sent back the pages to Hubbert for review, Hubbert's addendum had been deleted. The editor said Hubbert hadn't brought up the report during the meeting, when the AEC could have responded. The mention of suppression had been suppressed.

This deletion incensed another of the conference participants, a University of Minnesota professor of public affairs who passed on the information to Edmund Muskie, a powerful senator representing Maine. Muskie jumped on this issue, holding hearings of his Subcommittee on Air and Water Pollution. Muskie also asked Hubbert for permission to publish his addendum in the *Congressional Record* and formally requested the AEC to release the unpublished waste report.

Meanwhile, Senator Church continued hounding the AEC. "I am increasingly troubled over the trend toward secrecy in our Government," Church said in a Senate speech in March 1970. Regarding the waste committee's suppressed report, Church argued, "If security reasons are involved, or the Commission does not feel the report is factual, it should say so."

When a *New York Times* reporter wrote an article about Church's speech and asked the AEC for comment, the commission released the long-buried report to the *Times*. The newspaper noted the release as a brief aside—but it was a revelation to Hubbert and the other waste committee members. The committee's secretary told the other members they should consider it a de facto release—and should feel free to distribute the report themselves.

Over the coming weeks, the AEC sent out a handful of copies, but far more people wanted to read it. If there was enough interest, Hubbert told the committee, he'd pitch in to help pay to print hundreds of copies. When Senator Walter Mondale's office called Hubbert to ask for a copy, he told them, "I have only a single copy I'm defending with my life."

Energy Crisis

AFTER AN EXHAUSTIVE YEAR-LONG STUDY, in February 1970 Nixon's Cabinet Task Force issued a thick report, "The Oil Import Question." In reexamining the issue of oil import quotas, the report took input from the industry as well as the Interior Department about how much oil the nation might be able to produce in the coming years—and how much it would need to import.

As long as prices remained at their current level of about $3.30 a barrel, US production would continue increasing, the report forecast. To back that up, the report highlighted a forecast by Humble and its parent company, Standard Oil of New Jersey, which showed US oil production continuing to rise smoothly through 1980. The average of all the submissions, including from the Interior Department, came out about the same as Humble's. So the industry and government agreed that, barring a collapse of oil prices, there was no production peak in sight.

Although the report did consider the risk that OPEC might try to enforce another embargo or lower its production, the report argued this was not a major concern. For one thing, the study stated that the United States had a spare capacity of 2 million barrels per day, so it could quickly boost production by more than 20 percent in case of emergency.

The report recommended phasing out America's limits on oil imports and instead putting tariffs on imported oil. This would allow in more foreign oil, while still protecting the domestic oil industry. However US oil companies were struggling financially, and drilling had been declining continuously for the past fifteen years. Any proposals that might allow in more foreign oil were hugely unpopular with these companies—so Nixon ignored the task force's recommendations and kept policies as they were.

DESPITE THE CABINET Task Force's generally optimistic outlook for US oil production, there were reasons for worry. As US drilling declined, the nation's spare capacity eroded, and the Texas Railroad Commission continued increasing its allowables. During the year the task force had been compiling its report, Texas had raised the allowables drastically, from about 45 percent to nearly 70 percent. It was the first time since the early 1950s, when the United States was engaged in the Korean War, that the allowable had been so high. Yet no war or other conflict impinged on the world's oil production or trade. With US consumption soaring, there was a growing gap between the nation's consumption and its production—and increasing the allowables wasn't enough to fill that gap.

By the summer of 1970, there was increasing talk of an "energy crisis." Cities along the East Coast—New York, Philadelphia, even Washington, DC—suffered blackouts and brownouts. Chicago, St. Louis, and Minneapolis flirted with similar outages. The reasons for these shortages were many.

Oil had increased its share of the electricity market, rising to supply nearly one-seventh of the nation's electricity by 1970. To abide by laws to cut pollution, utilities had pushed aside coal in favor of fuel oil and natural gas—a boon to the petroleum industry. Meanwhile the nation's oil import quotas were being gradually relaxed and companies were exploiting loopholes—so oil imports were rising quickly. But still there wasn't enough fuel oil to go around.

Natural gas was another part of the problem. Despite decades of continual growth, production had recently fallen short of expectations. A major reason was the federal government's price controls on gas, put in place in the mid-1950s to keep the fuel affordable. By the 1960s, however, the controlled prices were so low that companies had been discouraged from searching for more gas. Also, proved reserves had recently fallen, suggesting the industry wasn't finding gas as fast as it was extracting it.

The troubles extended to other aspects of the electricity sector. Coal, the old stalwart, wasn't coming through. When oil and gas fell short of demand and utilities tried to raise their coal use again, they found that suppliers had already committed to selling their surplus coal overseas. Also nuclear hadn't lived up to its promise, hampered by chronic cost overruns and burgeoning public opposition.

As a result of all these forces, there simply wasn't enough energy—at least not in the right places. "We've given up hoping we can prevent a crisis," an official from the Federal Power Commission told a reporter. "Now we just hope we can stop a crisis from becoming a catastrophe."

THAT SUMMER OF 1970, US House of Representatives hearings on oil imports revisited the issues that Nixon's Cabinet Task Force had grappled with. When Humble's latest chairman of the board, Myron Wright, testified in these hearings, he presented the same outlook as his company had submitted to Nixon's task force, showing US oil production continuing to rise for at least another decade. But with consumption exceeding expectations and imports set to continue rising, Wright had a dire message: "We see a potentially disastrous petroleum supply situation."

For that same congressional hearing, Shell Oil submitted a statement with a stark title, "The Impending US Energy Shortage." Shell's forecast was lower than Humble's, arguing that "domestic production of petroleum, assuming continuation of import controls and the current real price levels, is expected to increase very little if at all." If the nation got rid of import controls, cheaper foreign oil would undermine the US industry's profitability—and the nation would be able to produce even less.

These industry views weren't quite the same as Hubbert's. But he was pleased to see they were no longer maintaining that the United States could supply all that its consumers wanted. As Hubbert wrote to a fellow researcher soon after that congressional hearing, "This is the first time to my knowledge that this situation has been publicly acknowledged by officials of major oil companies."

The following month, when Hubbert gave a public talk at a high school, he narrowed down his prediction for the timing of the oil peak. "We're somewhere right in this neighborhood of a peak right now, according to all the evidence that we now have," Hubbert said. "It may be this year, it may be next year, but we're in the neighborhood."

When oil data became available for 1970, it showed US production had set yet another new record, making a big jump from 9.2 to 9.6 million barrels a day, a 4 percent increase. Once again the industry had staved off the peak. However, to achieve this, Texas had continued ratcheting up the allowable oil production. By the end of 1970 allowables had surpassed 80 percent, leaving little spare capacity.

In July, the National Petroleum Council (NPC) released a major report, "Future Petroleum Provinces of the United States," which posited vast amounts of oil still available, estimating that the ultimate production for the contiguous United States could exceed 400 billion barrels—more than double what Hubbert thought likely. Nonetheless, given current policies and rates of drilling in recent years, the report expected that by 1985, production from the contiguous United States would decline by more than 1 million barrels per day—more than a 10 percent drop.

For the near term at least, much of the hope centered in Alaska. In 1968 came the discovery of the Prudhoe Bay field on Alaska's North Slope, holding some 10 billion barrels, the largest field that had been found to date outside the Middle East. This had kicked off a drilling frenzy in the frozen north. Although no more similarly large oil fields had been discovered in the couple of years since, hope for the state's potential soared, with estimates as high as 100 billion barrels. Alaska would keep US production rising, the NPC concluded, saying, "Were it not for the North Slope, the domestic trend would have turned down steeply."

Despite the optimism about Alaska's oil, the industry took on an increasingly grim outlook. After the NPC's report, a column in *Oil and Gas Journal* stated, "The possibility of an energy shortage within the next

10 years is very real, a fact which is being realized by more and more concerned experts every day."

In mid-1971, when Humble Oil issued its latest forecast for US oil production, it was "well below our earlier assessment," admitted Wright, the company's chairman. "We estimate our domestic petroleum supply will peak within the next two or three years, and slowly decline thereafter."

Forecasts from industry officials and petroleum press were at last approaching what Hubbert had been saying all along. But they didn't agree with Hubbert about why it was happening. "The crisis is being created," the *Oil and Gas Journal* column argued, "by an ever-increasing growth in energy demand plus restrictions on supply being imposed by new environmental regulations"—such as a ban on offshore drilling in California after a major oil leak.

Similarly, the NPC remained optimistic. Though it had presented an outlook for production to decline, it said this was not a forecast—it was instead a "projection," based on the assumption that the nation's policies would not change. NPC expected that the nation would change its policies—and then, "given adequate economic incentives," the industry could keep production rising.

Limits

"IT WOULD BE DEEPLY APPRECIATED if you could make available to us the services of Mr. M. King Hubbert, who would serve the committee for a few weeks as we gather statistics and information." This request from Senator Henry Jackson to the secretary of the interior, sent July 23, 1971, was innocuous enough. But at the Geological Survey, it set off a firestorm.

Jackson, a veteran senator from Washington state, was known as Congress's leading authority on a wide range of subjects—on defense, on atomic energy, on energy in general. Despite his authority and position, he went by his childhood nickname, "Scoop." He was painfully yet endearingly boring. (A joke about him: "Did you hear about the night Scoop Jackson gave a fireside chat? Halfway through, the fire fell asleep.") With his oversize head and stumpy arms, as one reporter put it, Jackson looked a hand puppet. His lack of charisma, though, belied his power. By 1971, he'd become a leading Democratic contender to challenge Nixon in the next presidential election. Scoop Jackson was a man to be listened to.

Jackson's Committee on Interior and Insular Affairs had recently launched a major energy study and was looking for advisers. Jackson's letter didn't ask the Interior Department to choose its favored expert from its staff, which included the Geological Survey. By directly requesting

Hubbert's help, Jackson broke the chain of command and rebuked the survey's expertise.

Just then the Geological Survey's leadership was in the midst of transition. The previous director, William Pecora, had left earlier that year, leaving the seat empty while the survey searched for a replacement. Next in line for the job, having recently become the survey's chief geologist, was Hubbert's longtime foe, Vincent McKelvey.

The survey's top men argued over whether to reject Jackson's request. Thomas Nolan, who'd served as director for a decade, was still around, having retired from administration to focus on research. Nolan gave his colleagues sage advice: if they wanted to stir up more problems, the best way would be to turn Scoop down.

Nearly three weeks after Scoop's request, Assistant Secretary of the Interior Hollis Dole finally approved it. Soon afterward, when Hubbert went for lunch in the Interior Department's cafeteria, he ran into Dole, who gave him a warning of sorts: "I signed that letter very reluctantly."

Several weeks later, Vincent McKelvey was nominated to become the Geological Survey's next director—and his confirmation would, in all likelihood, be a mere formality.

Just days after McKelvey's nomination, Hubbert got word that the survey was taking away his secretary, which would severely hamper his ability to respond to letters, write reports, and generally get things done. In a long memo Hubbert wrote for himself the day he received the news, he concluded, "I could hardly offer any other explanation of this action than a deliberate attempt to sabotage my assistance to the Committee."

IN JANUARY 1972, when the oil data became available for the previous year, it showed that, for the first time in a decade, US oil production had fallen. The drop was only slight, from 9.6 million barrels a day in 1970 to 9.5 million barrels a day in 1971. But with the nation fretting over the energy crisis and worried about rising oil imports, it was no time for domestic production to falter.

An *Oil and Gas Journal* headline cautioned "US Heading for a Close

Shave on Crude This Year." However, the article didn't mention that production had declined, nor the possibility that the United States had passed its all-time peak of production. Instead, it spoke vaguely of "slipping productive capacity." The decline in oil production failed to make headlines—both in the petroleum and the popular press.

But this decline was in line with what Hubbert had been predicting for the last fifteen years. His three-curve method—using statistics on discoveries, reserves, and production—had suggested the peak might come a bit earlier. But he'd also pointed out that if Texas used up its spare capacity, maxing out its fields, it would keep production rising a bit longer and delay the peak. Texas had indeed been raising its allowables, using up its spare capacity. And at that point, according to the Texas Railroad Commission calculations, the state still had some spare capacity left.

"AMERICA HAS ALWAYS been a growth society. But today an organized challenge is being mounted to that concept." That's how *Life* magazine put it in a 1970 spread covering a new movement called Zero Population Growth. The movement had been launched in part by two Stanford University biologists, Paul and Anne Ehrlich, who'd recently written an unlikely best seller, *The Population Bomb*, arguing that food production wouldn't keep up with population growth, so hundreds of millions worldwide would die of starvation in the decades to come.

Such advocacy of zero growth followed on the footsteps of the growing environmental movement, spurred in part by Rachel Carson's exposé on pesticides, *Silent Spring*. The United States had passed a variety of bills to protect nature and ecosystems, including Scoop Jackson's National Environmental Protection Act, which established the Environmental Protection Agency. Across the United States, there was a rising sense that there were too many people, consuming too much, creating too much pollution.

Zero Population Growth drew a line in the sand for one metric. Other movements focused on other metrics, such as zero economic growth. Boulder, Colorado, was a pioneer in putting boundaries around the city to try to limit sprawl. "The idea of a zero growth rate in the United States is becoming more respectable day by day," a newspaper columnist had noted in 1970. "A few years ago it would have been political dynamite. Now it can be discussed analytically without meeting an immediate and impenetrable emotional barrier."

In the early 1970s, a team of researchers at MIT had been investigating such limits to growth for the whole planet. Their study used a computer model developed by Jay Forrester, an MIT professor who had pioneered a field called system dynamics, using computer simulations to understand complex processes. (In the late 1950s, a graduate student of Forrester's had created one of the first such computer models, named SIMPLE—Simulation of Industrial Management Problems with Lots of Equations.)

Forrester's latest model, called World3, was a vast abstraction from the real world. It attempted to represent many major forces and the ways they could interact: the growth of population and industry, the consumption of food and natural resources, the buildup of pollution. The MIT team wanted to look at the relationships among these factors, to see if the world would run into a limit for one reason or another—and how the world might adjust. The team's work on this model had already received media coverage, the first article a long profile in an unlikely outlet, *Playboy*. "There it was," recalled Donella Meadows, one of the team members, "an analysis of population growth, economic growth, pollution, resource depletion—right there among the naked ladies."

Soon after the *Playboy* article, in late 1971, Hubbert received a letter from Dennis Meadows—another member of the team, and Donella's husband—saying their work had been strongly influenced by him. Hubbert met Meadows, then in his late twenties, and took a liking to him. Meadows suggested they collaborate. Hubbert was interested, but he had a lot on his plate and was worried about "administrative impediments." Probably the biggest hurdle was that he'd never been particularly good at working with anyone.

In early 1972, a few months after Meadows and Hubbert met, the MIT team completed a popularization of their results, describing the

World3 model and explaining its implications. They published it as a book, titled simply *The Limits to Growth*. Their conclusion: "physical limits to growth are likely to be encountered in the lifetime of our children"—that is, by the mid-twenty-first century, if not sooner. The results suggested that easing a limit on one factor, such as food, might allow a bit more growth, but soon another factor would become a limit. In scenario after scenario, continuing with business as usual and pushing for continual growth led the whole system to collapse—bringing a crash in industry, in food production, and in population. The study posed a vexing question: Would humanity anticipate such limits and create a stable society, or would it wait until it ran into such limits?

Funded by the Volkswagen Foundation, the team hired a public relations firm, which organized a symposium at the Smithsonian Institution—and though tickets were tight, Hubbert called around and was able to secure one.

Even before the symposium, articles about *The Limits to Growth* began to appear in the press. *The New York Times* gave it a generally favorable review, taking the idea of limits seriously even while quoting economists who were dubious about the notion of a steady-state society. "It's a simplistic kind of conclusion," said Simon Kuznets, a Nobel Prize–winning economist at Harvard. "You have problems, and you solve them by stopping all sources of change." Similarly, a Yale economist argued that a steady-state society would require people to be "very routine-minded, with no independent thought and very little freedom, each generation doing exactly what the last did." Neither of these economists, it seemed, could imagine that a society might face limits that would demand ingenuity merely to hold things steady—to repair wear and tear while also coping with depleting resources.

To counter such simplistic opposition to the notion of limits, the MIT team highlighted some of the benefits. "A society released from struggling with the many problems caused by growth," Dennis Meadows told the *Times*, "may have more energy and ingenuity available for solving other problems." People could devote more time to pursuits such as arts and music, education and religion.

The official launch of *The Limits to Growth* was held in the Smithsonian Institution's gothic castle on DC's Mall, where some 250 senators and representatives, ambassadors and academics, as well as Hubbert, gathered to hear Dennis Meadows describe the results. Under the television crews' bright lights, Meadows looked uncomfortable, but he gave clear explanations of the problems with exponential growth. The MIT team were not prophets of doom, Meadows insisted. They were optimists. They hoped that humanity would muster the will to make a smooth transition to a steady-state society.

The media coverage afterward—in *The New York Times Magazine* and *Washington Post*, *The Wall Street Journal* and *Newsweek*—critiqued the team's findings and conclusions. A variety of experts, principally economists, argued against the study. It was a computer model, they said. It was too simplistic. It hadn't taken economic factors, such as rising prices, into account. Some argued growth was indispensable, the only way to pull people out of poverty.

Anthony Lewis, a respected *New York Times* columnist, was a rare voice of strong support of the study. "Growth is a cop-out, a way of avoiding the real social and moral issue of equality," Lewis wrote. "Facing the ecological truth about our planet should help us face those issues at the same time."

In talks soon after the release of *The Limits to Growth*, Hubbert noted he'd been saying much the same thing for decades—but he wasn't bitter. He was encouraged to finally see the nation increasingly questioning growth. For the interior secretary, Hubbert wrote a long commentary commending the MIT study. In every one of the study's scenarios that pushed growth as long as possible, each "reached some kind of crisis within a century or less," Hubbert explained. "The only solutions that do not lead to a crisis are those that are based on a transition at an early date from present exponential growth into a stabilized state."

Hubbert saw *The Limits to Growth*, and the debate it generated, as a turning point. He concluded, "It would not be surprising if, when viewed in retrospect of a few decades hence, this may prove to have been the beginning of one of the greater intellectual liberations and advances of the present century."

WEEKS AFTER THE symposium on *The Limits to Growth*, the United States hit a limit of its own. In March 1972, the Texas Railroad Commission raised the allowable oil production yet again, this time from 86 percent to 100 percent. "Texas in for Year-Round, All-Out Production," read a *Dallas Morning News* headline. In one stroke, America's spare capacity was officially gone. Barring some huge, unforeseen discoveries, the state's oil production was at its peak.

Even when producing flat out, Texas would still yield less oil than hoped for, explained Byron Tunnell, the commission's chairman. The spare capacity had actually eroded in recent years, so the official figures didn't reflect reality. The jump in the allowable to 100 percent wouldn't actually boost production much, Tunnell announced, adding, "Maybe that will get the message to Washington."

In case anyone missed the geopolitical implications, Tunnell spelled them out. Texas yielded more oil than any other state, and with Texas going all out, the United States could no longer quickly ramp up production in an emergency. The nation would become increasingly dependent on foreign oil. And the exporters—chiefly members of OPEC—would recognize their newfound power and demand much higher oil prices. As Tunnell put it, OPEC would extract its "pound of flesh."

On reading this news about the end of America's spare capacity, Kenneth Deffeyes, Hubbert's former colleague at Shell, thought to himself, "Old Hubbert was right."

The Republican Ethic

THE TEXAS RAILROAD COMMISSION'S MESSAGE may have gotten through to Washington—but it was soon eclipsed.

In June 1972, five burglars were arrested in the middle of the night inside the Watergate office building, caught in the act of trying to install listening devices—bugs—in the Democratic National Committee offices. The burglars were part of a secret White House intelligence unit that Nixon had authorized to use any means necessary, including illegal acts, to collect information on rivals.

John Ehrlichman—Nixon's chief domestic adviser, as well as the White House's top adviser on energy—had been one of the main orchestrators of the break-in. It was only the tip of what he and others in the White House had been involved in. Ehrlichman had also ordered a break-in at a psychiatrist's office, for example, to look for damning information about Daniel Ellsberg, a military analyst who had leaked a top-secret report on Vietnam that became known as the Pentagon Papers.

When the Watergate burglars were caught, suddenly Nixon, Ehrlichman, and the whole administration had much more immediate problems on their hands.

. . .

"THINGS HERE ARE buzzing around so fast that I cannot keep up with them," Hubbert wrote in September 1972 to David Willis, his former assistant at Shell.

Hubbert had recently talked with a reporter at *Fortune* magazine. The head of the US Federal Power Commission had favorably cited Hubbert's forecasts. Hubbert had also become friends with Stewart Udall, the interior secretary under Kennedy and Johnson. Although Udall had been in office while Hubbert was issuing his warnings, Udall hadn't heard them. The two of them met only after Udall left office, when they served together on the board of trustees of a nonprofit, the Population Reference Bureau, that warned about overpopulation. Udall was bitter about having been misled by the Geological Survey's oil and gas estimates. To try to make amends, Udall spoke out and wrote articles, including a long piece in *The Atlantic Monthly* that called Hubbert "the one petroleum expert who has unerringly forecast the curve of our domestic production." As Hubbert told Willis, "After all these years, I seem to be getting recognition as some sort of a prophet."

The following week *Fortune*'s latest issue had a cover story on energy resources, and to promote it the magazine ran a full-page advertisement in *The New York Times*. A bold line occupied most of the page, curving up the left side, reaching a pinnacle at the top, and then descending on the right side of the page. Underneath, in large font, it said: "What on earth can we do about Hubbert's Pimple?"

AS THE 1972 presidential election drew closer, Nixon was worried about the energy outlook. Although he didn't say much about it publicly, he told one of his top aides, "It should scare the hell out of people."

A few years earlier one of the Nixon administration's first major efforts was its Cabinet Task Force on Oil Imports—but the president wound up ignoring the report's recommendations and making no major energy policy changes. Since that time, his administration had been consumed with trying to unwind the Vietnam War, widely called a "quagmire" and a "morass," while also attempting to stymie investigators who were discovering that the Watergate burglars had ties to senior members of the Nixon administration.

Meanwhile the nation was caught in an economic malaise dubbed "stagflation," in which economic growth was low, yet inflation was high, eroding consumers' purchasing power. This situation—believed to be impossible, according to leading schools of economic thought—came to be synonymous with the administration's economic policies, labeled "Nixonomics."

On top of these issues, the administration was extremely busy with its campaign for reelection. Despite the various headwinds—economic, military, and social—as well as the Watergate scandal, Nixon easily won the election.

With the election over, the question of oil supplies loomed larger. Hollis Dole—the assistant secretary of the interior who'd only reluctantly granted approval for Hubbert to help Scoop Jackson—warned of the nation's "rapidly increasing dependence on imported petroleum." Dole requested the National Petroleum Council to prepare a new report on the issue—and to report back quickly, in only six months. The secretary of commerce also made a bold proclamation at a meeting of the American Petroleum Institute: "The era of low-cost energy is almost dead. Popeye is running out of cheap spinach."

As new data came in, America's oil outlook looked increasingly dire. When the data for 1972 appeared in *Oil and Gas Journal*, it showed US oil production had dropped again—the second year in a row. For years, imports had made up about one-quarter of US oil consumption. But following the 1970 peak of US production, in just two years imports had soared to more than one-third of the nation's consumption.

The outlook for natural gas was turning sour as well. Proved reserves of natural gas had plateaued in the late 1960s—and the latest data showed they dropped sharply in both 1971 and 1972. If Hubbert's methods were right, a decline in natural gas production was also imminent.

. . .

IN THE SPRING of 1973, Hubbert headed off to Berkeley to spend a quarter at the University of California as a regents professor, a visiting position for prestigious scholars. While at Berkeley, Hubbert wasn't required to do much of anything, but he gave dozens of talks at seminars around campus. He joked to William Rubey, "They nearly worked me to death." But really it was his own doing. He accepted as many talks as he could—and by then, as a profile in the *San Francisco Chronicle* put it, "Hubbert is one of the most sought-after lecturers in the nation in this ecology-conscious, energy-crisis-ridden age."

In a talk titled "Social Implications of a Finite World," which Hubbert gave to graduate students in social sciences and law, he riffed on the chemist C. P. Snow's idea that science and art were "two cultures" separated by a wide gulf. Hubbert was concerned with a different, more momentous division: between science, which dealt with matter and energy, and "the monetary culture," which measured everything in terms of dollars. Ignoring the physical world, followers of this monetary view assumed that growth could continue indefinitely.

An exemplar of focusing on money rather than material things was Jesse Jones, who'd been in charge of stockpiling rubber during World War II. Japan had cut off US access to the main rubber supplies from Southeast Asia, making America's limited supplies practically priceless, since no amount of money would allow the nation to buy more. But Jones didn't seem to grasp this. As Hubbert told the story, when a warehouse full of rubber burned up, Jones said, "It was insured, wasn't it?"

Hubbert's stay in Berkeley was a time of tumult. Revelations of secret bombing campaigns in Cambodia spurred more protests over the Vietnam War, and Nixon's cover-up of Watergate was unraveling. The public was increasingly skeptical about the government—including its assurances about energy.

When Hubbert gave a series of three public talks held on Friday evenings on the Berkeley campus, titled "Energy and Man," each lecture drew audiences of four to five hundred. Overall the experience at Berkeley was "strenuous and intellectually exciting," Hubbert told Stewart Udall. "I have never before experienced such complete acceptance."

Describing the situation to his old friend Wallace Pratt, Hubbert wrote:

> There is a very great intellectual ferment going on in the country in all major categories—government, industry, and academic—over just what is the state the world has got into, and how are we going to get ourselves out of it. Certainly, there is more competent social thinking being done—more than at any time since the depression, and I am disposed to think that it greatly surpasses that.

SOON AFTER NIXON'S second inauguration, his administration made a new push to tackle energy issues. In early 1973 it brought on board James Akins, the State Department oil expert, to lead a major new effort: crafting the nation's first comprehensive energy policy.

By then, Akins had gained a reputation in DC circles as a crusader for conservation. He was a calm man, a Quaker—but also self-righteous. To reach his office each day, he left at six a.m., walking three and a half miles. To save energy, he turned down his thermostat in the winter.

Akins did much of the initial drafting of the new energy policy, but rifts soon broke out. Ehrlichman told the newcomer: "Mr. Akins, you've got to understand that conservation is not the Republican ethic." (Or as Ehrlichman put it on another occasion: "This will not be the administration to tell people to turn down their thermostats!") Similarly, Nixon and his top aides—including Henry Kissinger, his national security adviser—downplayed OPEC's power. Nixon eventually took the policy work away from Akins, passing it to Kissinger and a small team of other administration stalwarts. They had little experience with oil but were known for their skill at writing speeches and policies in "Nixonese."

Akins then struck out on his own. In March 1973, in a long article in the influential journal *Foreign Affairs*, he laid out his views on the energy crisis. He pointed out that Arab nations had issued many warnings over the previous several years. Earlier that year, for example, the Kuwait National Assembly had voted unanimously to use oil as a political weapon against Israel, and the president of Syria had said, "We can now use Arab oil in all our battles against our imperialist enemies." Most in the US government dismissed these warnings, Akins wrote, with arguments like "They need us as much as we need them" and "They can't drink the oil."

However, "the threat to use oil as a political weapon must be taken seriously," Akins argued. "The vulnerability of the advanced countries is too great and too plainly evident," in large part because "the United States now has no spare capacity."

Akins admitted that attempts to predict the future of oil had a dismal record, with "only marginally better success than those who foretell the advent of earthquakes or the second coming of the Messiah." He conceded that in the past there had been many false alarms. "Oil shortages were predicted in the 1920s, again in the late 1930s, and after the Second World War," he wrote. "None occurred, and supply forecasters went to the other extreme: past predictions of shortages had been wrong, they reasoned, therefore all such predictions must be wrong and we could count on an ample supply of oil for as long as we would need it."

But by 1973, the situation was different, Akins argued. As the title of his article put it: "This Time the Wolf Is Here."

SAUDI ARABIA—THE LEADING member of OPEC, and the largest oil exporter in the world—was acutely aware of America's predicament: that its oil production was falling, its spare capacity was gone. OPEC had been ratcheting prices up, forcing Western companies that extracted the oil to pay more per barrel—costs they passed along to consumers.

To fight OPEC, some US senators talked of consuming nations banding together, and the State Department began negotiations with Europe and Japan to form such a "buyer's cartel." In response, Saudi Arabia's oil minister said, "If it is war they want, then war they shall have. They will be starting it and they will be the losers."

When Nixon announced his administration's new energy policy, in a special message to Congress on April 18, he addressed growing worries about oil supplies. "We should not be misled into pessimistic predictions of an energy disaster," Nixon said. "But neither should we be lulled into a false sense of security."

Nixon claimed America could produce more fuels, but his pronouncements were vague. "We have potential resources of billions of barrels of recoverable oil," he said, and "similar quantities" of oil shale. Nixon argued these resources were more than adequate. "Properly managed, and with more attention on the part of consumers to the conservation of energy, these supplies can last for as long as our economy depends on conventional fuels."

Nixon called for saving energy, in a strategy that balanced the views of Akins and Ehrlichman. "We as a nation must develop a national energy conservation ethic," the president said. However, he argued that conservation "can be undertaken most effectively on a voluntary basis." His administration would ask nicely but wouldn't force anyone to turn down their thermostats. Its main emphasis was on boosting supplies.

Nixon referred only obliquely to America's falling oil production. "We are not producing as much oil as we are using," Nixon said. "We are being forced to obtain almost 30 percent of our oil from foreign sources."

But rather than see these imports as a major risk, Nixon opened the floodgates to foreign oil with a major change of policy: he would erase Eisenhower's import quota system. The new policy, Nixon said, "should help to meet our immediate energy needs by encouraging importation of foreign oil at the lowest cost to consumers." He made no mention of where that foreign oil came from.

After September 1, 1973, however, Nixon did begin talking more frankly about oil imports. That was the day Libya's leader, Colonel Muammar Gaddafi, nationalized his country's oil fields, wresting control from international companies. At that point, Gaddafi had already pressured a few oil companies into paying Libya more for each barrel extracted. With nationalization complete, he hiked the oil price 30 percent, from $4.60 to $6 a barrel. A few days after Gaddafi's move, Nixon publicly acknowledged the links between America's oil imports and its Middle East policies. If Congress did not act on his energy policy proposals, Nixon said, "it means we will be at the mercy of the producers of oil in the Mideast."

But at the same time, the president warned OPEC. "Oil without a market does not do a country much good," Nixon proclaimed. "We and Europe are the market."

HUBBERT CONTINUED TO work on his own, but the Geological Survey tried to get him to help with various projects, including an evaluation of "nuclear fracturing," a technique being tested by the AEC as part of its Project Plowshare.

In 1967 the commission had done a test in New Mexico called Project Gasbuggy, setting off a nuclear bomb underground in a shale layer to try to fracture it and release more natural gas. It had done two further tests like this to unlock natural gas, the most recent earlier in 1973, and it was looking to do more.

Such efforts were seen crucial because natural gas reserves were falling, and production had plateaued. It appeared US natural gas production had reached its peak—albeit a few years sooner than Hubbert had predicted.

Nuclear fracking, the AEC claimed, could double the nation's natural gas reserves. Hubbert was intensely skeptical about such claims, and about the whole approach. For one thing, the gas that came out was radioactive. The AEC and other boosters said the radioactivity was low, so it would add only 1 percent to the exposure people would receive during everyday life.

But the AEC response to such concerns had been "pitiful," wrote Hubbert's boss, Charles Masters, in a memo to Hubbert. "Somehow the AEC must be goaded into open scientific debate—the stakes are too high to let legitimate scientific concerns go unanswered," Masters said. He urged Hubbert to study this issue and help the Geological Survey "develop an in-house opinion on the status of nuclear frac." Masters added, "I know of no other man in the country more capable than yourself of rigorously analyzing the situation."

"The study you propose needs to be done, and I am happy to do it," Hubbert replied to Masters. However, he lacked an assistant or a secretary, and he was already working incessantly. "The handicap is that I can muster only a limited number of manhours per 7-day week."

Off and on, for the past couple of years, Hubbert had continued to work on a report for Senator Henry Jackson. Initially asked simply to advise the senator for a few weeks, Hubbert had made arrangements with the senator's office to write a long report as an update to his reports for the National Academy of Sciences, published in 1962 and 1969. By the fall of 1973, Hubbert was finally wrapping up a report for Jackson, which had ballooned to the length of a book, covering oil and gas with a bit on coal—and that was just the first part. He still intended to write a second part to cover all other energy sources, including nuclear, solar, and hydroelectric.

By the time he was writing his report, it was clear that US oil production had reached a peak in 1970 and then declined sharply—a decline that Hubbert expected to continue. It looked to Hubbert as if the peak of US natural gas production was likewise imminent.

Many had hope for Alaskan oil and natural gas to make up for declining production in the rest of the nation. There was a proposal for a pipeline that would cut across Alaska, bringing oil from the Prudhoe Bay field, on the especially harsh North Slope, to a more accessible southern port at Valdez. The proposal had gone through two years of assessments of the pipeline's likely environmental impact, and environmentalists had gone to court to try to stop construction. But by the summer of 1973, Congress blocked further lawsuits over the pipeline. So five years after the discovery of Prudhoe Bay, the nation's largest oil field, it looked as if it would soon begin to yield fuel.

In his Senate report, Hubbert allowed that there might be more big finds in Alaska, so the whole state might ultimately yield some 40 billion barrels—four times the amount in the Prudhoe Bay field. Still, he didn't expect such a quantity to significantly change the nation's dependence on foreign oil.

For decades, US officials and the petroleum industry alike had maintained that if conventional oil fell short, it could be replaced with oil shale. With US oil production declining, there was increasing talk about oil shale, with some proposing to start extracting it in Colorado, scaling up to 100,000 barrels a day by 1985. But that quantity wouldn't make much of a difference, Hubbert argued in his report. To make any significant dent in America's dependence on foreign oil, they'd need to scale up at least ten times more, to 1 million barrels a day. That level, however, would require mining about 1.25 million tons of shale per day—nearly as large a quantity as the nation extracted daily from all its coal mines. Then the industry would have to cook the oil shale to break it down to create liquids that could then be refined into fuels like gasoline.

In this process—crushing the rock and having it react with oxygen— the subsequent wastes actually occupy a larger volume than the rock extracted from the ground. Most of the wastes could be buried in the same mines that the shale had been extracted from, but there would still be leftover wastes that wouldn't fit. Some had proposed dumping these wastes into canyons near the oil shale deposits. But both methods of waste disposal—underground and in canyons—had problems, Hubbert argued. As rainwater and groundwater seeped through the wastes, it would likely leach out polluting compounds such as calcium oxide and magnesium oxide, which could foul the region's rivers.

What's more, the process of getting the oil out of the shale would require about ten barrels of water for every barrel of oil created. There likely wouldn't be enough water from the Colorado River, since most of the available water had already been allocated to farms and cities.

Large-scale oil shale mining, Hubbert concluded, would "approach an environmental calamity."

Then there was the question of how much oil shale was really feasible to extract. In the mid-1960s, the Geological Survey had estimated there were some 600 billion barrels recoverable in the Green River formation, underlying Colorado, Utah, and Wyoming—but only about 80 billion barrels of that appeared economical to mine. More detailed work by the National Petroleum Council in 1972 found that most of the Green River oil shale was too low grade to be practical to extract, leaving only 20 billion barrels that might be economic to extract and cook down. That amount would add only about 10 percent to the nation's ultimate liquid fuel supply. It might help somewhat but wouldn't substantially change the overall outlook, Hubbert argued.

After Hubbert's many conflicts with the Geological Survey, he didn't trust them to handle any aspect of his Senate report. Miriam typed up the whole report for him at home, even as it mounted to over two hundred pages. Because events were roughly following Hubbert's expectations, his report contained no bombshells. It was also too long, too technical, and too late.

Since Scoop Jackson had requested Hubbert's help, the senator had made a run at the White House, which found only middling success, and he had quit his campaign. Then in the spring of 1973, Jackson had held hearings on oil—but his focus was on whether the energy crisis had been a "premeditated plan" of the oil majors to squeeze out the smaller independents.

After Hubbert turned in his report, Scoop Jackson's office sat on it.

AS THE ENERGY crisis deepened, Hubbert continued with his many talks. In a public lecture at Stanford in early October 1973, he argued the nation's problems were just symptoms of longer trends: "Our institutions, our system of accounting, our monetary system, our legal system, our government—the whole works—are premised on a continuing exponential growth." However, this growth phase, at least for countries like the United States, was "just about over," he said. "Now we're running into a situation where we cannot keep up exponential growth."

Hubbert explained he didn't want to create a sense of alarm or doom. He still was hopeful about humanity's ability to cope with the situation. The problem wasn't so much about resources or technology but about ways of thinking. "We're going into a cultural shock or crisis far more than we are going into an energy crisis," he concluded.

Very soon, though, the situation would grow far worse.

PART 5

Washington, DC, 1973–1989

180 Degrees

"THE SAUDIS ARE GETTING HEADY over the power of oil," Secretary of Defense James Schlesinger told other top officials, including Secretary of State Henry Kissinger and the head of the Central Intelligence Agency (CIA). They met over breakfast in the White House's Map Room on November 3, 1973, just weeks after war had broken out again in the Middle East, and Saudi Arabia and other members of OPEC had slashed their production and instituted an embargo against the United States.

These top US officials were in the midst of developing a plan for the Marines to invade the small, oil-rich kingdom of Abu Dhabi on the Arabian Peninsula and seize control of its oil fields—and to send a warning to other OPEC members. Even before the embargo, Marines had been training in the Mojave Desert for such an invasion. Soon the US military would have two navy destroyers at the entrance to the Persian Gulf, and an aircraft carrier, the *USS John Hancock*, was moving into the area.

"We need a public line on the *Hancock* when it arrives," Schlesinger said.

"Routine," Kissinger said. "An exercise we have been planning a long time."

After discussing strategies for the conflict between Egypt and Israel, and how to keep the Soviets from getting too involved, Kissinger concluded, "Let's work out a plan for grabbing some Middle East oil if we want."

"Abu Dhabi would give us what we want," Schlesinger replied.

DURING THE 1956 Suez Canal crisis, Saudi Arabia had instituted an embargo, to little effect. In 1967, OPEC had tried to use an oil embargo as a weapon, and that likewise fizzled. Throughout 1973, OPEC members—including Saudi Arabia, Libya, Iraq, and Kuwait—had been again warning that if the United States didn't change its policies toward Israel, they would cut off oil exports to America. Most US officials did not take the warnings seriously.

The situation in 1973, however, was much different. The United States was still the world's largest oil producer—but it was also the world's largest oil consumer. The nation had eaten through its spare capacity. Its production was falling while its consumption continued rising. As James Akins, the White House energy adviser, had predicted several months earlier, "This time the wolf is here."

Egypt's leader Anwar Sadat had wanted to break a stalemate following Egypt's 1967 war with Israel, and to push the United States to be more even-handed in its policies toward the region. After tensions in the region mounted for months, in early October 1973 Egypt launched a surprise attack on Israel.

In response, United States tried to placate Arab nations, while also quietly aiding its longtime ally Israel. But when Arab nations discovered this aid to Israel, they followed through on their threats, slashing oil production by 10 percent and vowing further cuts until the war was resolved, and hiked prices by nearly double. They also instituted an embargo, barring tankers from carrying their oil to the United States. International oil companies followed the embargo to the letter but undercut the spirit of it by reshuffling global oil shipments—just as they had during earlier embargoes. But this time, OPEC had more power over world markets. They were able to restrict the total amount of oil for sale, so consumers had to swallow higher prices.

Up to then, the Nixon administration had done almost nothing to prepare for such a situation. Nixon's top foreign policy adviser, Henry Kissinger—by then promoted to secretary of state—had been busy trying to negotiate a cease-fire in Vietnam, for which he won a Nobel Peace Prize earlier in 1973. Others in the administration had been consumed with reelection, continuing stagflation, and the deepening Watergate scandal.

Once OPEC instituted the embargo, the Nixon administration's outlook suddenly flipped. Kissinger considered it "blackmail" and pressured the major oil-consuming nations to respond by forming a united front. Meanwhile Kissinger and Schlesinger continued discussing plans to invade Abu Dhabi.

On November 11, Nixon took to television and gave the American people a stark warning: "We are heading toward the most acute shortages of energy since World War II." Nonetheless he remained positive and reassuring. "This does not mean that we are going to run out of gasoline or that we will freeze in our homes," he said. "The fuel crisis need not mean genuine suffering for any American. But it will require some sacrifice by all Americans."

For the longer term, Nixon had a vision. Invoking the Apollo Project to put a man on the moon, and the Manhattan Project to develop the atomic bomb, he called for a new national goal of freeing America from "foreign energy sources"—which meant oil, the only energy source the United States imported in any significant quantity. "Let us pledge," he said, "that by 1980, under Project Independence, we shall be able to meet America's energy needs from America's own energy resources."

THE DAY AFTER Nixon's speech, Hubbert began a long lecture tour—a "man killer," he called it, nine weeks of travel across the United States and Canada. The tour was sponsored by the American Association of Petroleum Geologists, which had invited him to make the tour months before the OPEC embargo. Hubbert's talk, "The World's Energy Economy," covered his overall outlook—in particular, the peaks of US and world oil and the coming end of growth. He typically drew 150 or 200 people at a time, over the whole tour reaching some ten thousand people.

"Audiences are no longer disposed to argue," Hubbert told Stewart Udall, the former interior secretary, when they spoke in the midst of his tour. "There is a very sober, thoughtful attitude with regard to the situation."

Just as audiences had changed their attitude, Hubbert told them he'd changed his mind about a crucial issue: the world's ideal energy source for the long term.

Throughout his time advising the Atomic Energy Commission, Hubbert had been critical of its handling of nuclear wastes. All along, he had thought these problems could be fixed—and must be fixed—because he had considered atomic energy essential for maintaining industrial civilization for centuries or millennia. He'd followed the technology as it emerged from military applications—for building bombs and powering submarines—and became a commercial reality. He'd seen nuclear power generation rise quickly through the 1960s and early 1970s, so that by 1974 nuclear supplied 6 percent of US electricity—a significant portion, but still less than the share of electricity derived from hydroelectric dams or natural gas, or even from oil.

After advising on nuclear power for more than fifteen years, Hubbert was frustrated. He felt the AEC showed little interest in handling nuclear waste carefully. Breeder reactors, which he thought essential, had been treated as a side-project. Nonetheless, as recently as 1972 he'd stated, in an interview with *Newsweek*, that nuclear was "the only source to meet the world's power requirements in the future."

By the time of his AAPG lecture tour in 1973, he told audiences, his view had turned around "180 degrees."

"Fifteen years ago I was like everyone else in thinking nuclear power would help meet our energy needs," Hubbert said in one talk. "But I've gradually come around to look at the hazardous aspects and it scares the hell out of me." The same technologies for creating nuclear power plants could be used to assemble the material for more warheads, he pointed out—and in "this unsteady world, with a propensity for throwing bombs at each other, the chances of a nuclear disaster have become increasingly frightening." It wasn't simply the Cold War face-off between the Americans and Soviets. Terrorists might attempt to "hold up New York or London or Paris," he argued. "It doesn't take any more courage to hijack a power plant than it does to hijack an airplane." He concluded that nuclear was a "perpetual hazard," creating wastes requiring "perpetual care."

In the early 1960s, Hubbert had thought solar power might be feasible only for developing countries or for special applications such as satellites. By the early 1970s, he'd come to see more promise in this approach, arguing in favor of government-funded research on large-scale solar power plants. He thought people had the basic knowledge to build them but warned that "the technological difficulties of doing so should not be minimized." Meanwhile he'd continued following solar research and development—attending international conferences and visiting with scientists to learn of their progress.

With Hubbert's 180-degree turn, he touted the power of the sun. "Solar energy dwarfs everything else in sight," he argued. "It turns out the big source of energy on this earth is sunshine. It's inexhaustible. It's been pouring in for billions of years and will continue for billions of years when the human species isn't here."

Though he'd downplayed solar power before, he admitted, "I'm happy to say that I was proven wrong."

AS HUBBERT TRAVELED on his lecture tour, the Nixon administration continued with efforts to resolve the Middle East conflict. Kissinger and Schlesinger discussed how to get Congress to approve a $2 billion package for military aid to Israel, and how to break OPEC's embargo.

In a meeting in the White House's Map Room on November 29, six weeks after the start of the embargo, Kissinger said, "The Saudis are blinking."

"They think we knocked off Idris," Schlesinger said, referring to the king of Libya, who'd been deposed by Muammar Gaddafi a few years earlier.

"They have never played in this league before. They are scared," Kissinger said.

"We need to build a presence in the Middle East," Schlesinger said.

"It is essential," Kissinger replied, adding, "Can't we overthrow one of the sheikhs just to show that we can do it?"

Meanwhile consumers around the world felt the bite of OPEC's oil cutbacks. In the eastern United States, there were local fuel shortages. Hit first and hardest were independent truckers, whose earnings were slashed once oil prices shot up. They began blockading freeways and turnpikes from New Jersey to Ohio and planned nationwide protests. One trucker told a reporter, "We want Nixon and his people, when they turn on the television, to hear us."

In some areas, gas stations ran out of fuel after being open only a few hours each day. Several states instituted "odd-even" rules, by which people had to take turns, able to fuel up only every other day, depending on whether their license plate ended in an odd or an even number. One Pennsylvania gas station attendant recalled that when tankers drove through town on their way to deliver more fuel, "motorists would follow the trucks right up to the pumps."

To try to manage the situation, Nixon created a Federal Energy Office and put his treasury secretary, William Simon, in charge. When asked when the United States would consider national fuel rationing, Simon replied, "I would say a critical factor would be if people begin queuing up at gas stations for three or four hours at a time." *The Boston Globe* commented, "A government policy based solely on visible chaos is something to ponder."

In the face of shortages, many called for the United States to boost its oil production. The development of Alaska's Prudhoe Bay oil field—estimated to be the nation's largest discovery to date—had been stuck in legal limbo for four years. Within a month of the OPEC embargo, however, Congress overrode the legal challenges to the pipeline across Alaska, finally pushing through its approval. However, it was a "triumph of scare propaganda and economic pressure over reasoned public policy," argued a *New York Times* editorial, since the pipeline would take years to build, providing no help in the short run.

Before the embargo, the United States had already been suffering from stagflation. With the oil price hike, inflation grew worse, eroding consumers' purchasing power, and the economy sank into a recession. The price hike had an even larger effect on other nations entirely dependent on oil imports—all of western Europe, and fast-growing Asian countries such as Japan and Korea.

In the face of higher prices for gasoline, job losses, and a worsening economy, Americans cut back their consumption. When they did buy cars, they opted for more efficient ones, leading *Time* magazine to declare on its cover, "The Big Car: The End of the Affair." Speed limits in many locales were lowered to fifty miles per hour, and some states gave incentives to employees to carpool to work. Businesses and homes lowered their thermostats and turned off unnecessary lighting, including advertisements and streetlights. The floodlights that had lit Chicago's Wrigley Building almost continuously for half a century were switched off.

By early 1974, OPEC members talked openly of lifting the embargo and raising production. Anticipating an end to fuel shortages, a top energy official announced that the federal government would sell off oil in emergency stockpiles, "so that we can release the energy supplies needed to support sustained economic growth." The government was eager to get back to normal—that is, to times of growth.

In mid-March 1974, most OPEC members—including the biggest producer, Saudi Arabia—did end their embargoes. However, oil prices remained as high as ever, which took a heavy toll on the US economy. With Americans questioning their addiction to cars and their assumption that resources were essentially limitless, the oil shock turned out to be, as Hubbert had put it, a "cultural shock."

This Sacred Thing of Growth

BY 1974, HUBBERT HAD SPENT a decade in Washington, where he'd moved in hopes of having "national influence." But beyond Senator Jackson's request for a report, Hubbert's forecasts had received little interest from politicians. On April 19, though, he received a call from a staffer for the US Senate's powerful Commerce Committee, led by Warren Magnuson, a member of the exclusive "inner club" that essentially ran the Senate. Having heard that Hubbert had long ago predicted the energy crisis, the staffer said, "We'd like you to testify before the committee." The hearing would begin in just ten days.

"Well, I'll look in my little black book here and see if I'm clear that date," Hubbert replied.

"Never mind your little black book," the staffer shot back. "You'll be here!"

He got the message. "I'm very happy to testify before this committee," he said. "But I want this understood. I'm not going to agree with the Geological Survey, so if the Geological Survey attempts to interfere with what I am telling you, I will not testify."

Magnuson's office understood his position, and it wasn't a problem. However, the survey informed Hubbert it would be impossible to give him approval to testify by the first day of the hearings. Hubbert waited.

Meanwhile the hearings went ahead. The topic was the Domestic

Supply Information Act, a bill Senator Magnuson had cosponsored, aimed at gathering better data for government analysis of supplies of energy and critical materials. As the senator put it, the aim was to "prevent or defuse the impact of future shortages."

On April 29, Magnuson opened the hearing with a speech about America's current plight, saying:

> We have never before this administration experienced simultaneously the twin specters of inflation and recession of this magnitude and duration. Never before have the people of this country been faced with the daily threat of both lost jobs and the incredible shrinking purchasing power of the dollar. These plagues seem to have descended upon us without warning. But we now know that these shortages were predictable. We have slowly come to the realization that we should have begun energy and resource conservation and other appropriate responses years ago.

Hubbert eventually received approval to testify in early May. In granting permission, McKelvey reiterated that Hubbert needed to make clear, in both his spoken and written statements, that he was not representing the Geological Survey.

In his testimony on May 10, Hubbert recapped his record of forecasts, and argued the government should do the same kind of work, establishing an elite group of researchers that would track historical data and forecast the future. "That can very definitely be done," he said. "I have been doing it very effectively for the last twenty-five years."

If the government were to form a new body for developing reliable forecasts, Hubbert argued, it would have to be willing to present bad news. Since ancient times, people in such a position "have frequently been executed," he said. "We don't do that so crudely now but still, psychologically, we don't like bearers of bad tidings." To enable such a group to do their job well, they needed to be insulated from administrative pressures, as well from influence by industry and politicians—anyone who "might prefer to have some other news than what the data indicate."

However, bad tidings, even when backed up with evidence, often went ignored, Hubbert pointed out. For example, his 1962 report for the National Academy of Sciences, despite having been commissioned by President Kennedy, had attracted little interest at the time. It came into great demand only after the nation found itself in an energy crisis.

When Hubbert finished, the first question came from Senator John Tunney, a California Democrat with an engaging, long-toothed smile. The coauthor of the landmark Endangered Species Act, Tunney had also written the bill under consideration, which was in line with his general goals of sidelining special interests and advocating long-term planning. Referring to Hubbert's 1962 academy report, Tunney asked, "What is it about the bureaucratic structure that makes such a report lie on the table collecting dust without decision makers paying attention to it?"

"That is a problem with which I have been personally very deeply concerned over the years," Hubbert replied. Some of the strongest objections he'd received were from the Geological Survey. Like most government agencies, he argued, the survey strived to justify its existence and maximize its budget. Hubbert described the survey's attitude as "We have to try to show that we are doing a good job, we are very indispensable to the national welfare. We can best do that by saying there are large amounts of resources and we can find them for you." Hubbert added, "There is a kind of a psychological drive to do this."

"Do you feel," Tunney asked, "that we need some separate agency that is responsible for evaluating material shortages, and not only forecasting, but additionally acting as an advocate as to what remedial measures are necessary?"

"I think perhaps you have touched on the critical point here," Hubbert replied. If this new body were simply part of an existing agency, that could create "an internal conflict of interest that could be very, very serious." The National Petroleum Council—an organization independent of policy makers but composed of industry representatives—was not a good model, either, he argued.

Hubbert then highlighted what he felt was a basic misconception in the Domestic Supply Information Act. In aiming to collect more and better information, the bill implied the problem was a lack of access to data. However, Hubbert argued, "we have enormous quantities of information already in existence within the various data collecting bodies of the Government. Most data is useless. We can bury ourselves in data every week. We do it all the time." Far more important was "having a fundamental knowledge of the problem being investigated," he said. If the government created a new agency for tracking resources, "the higher the competence of this staff, the more sense they can make of the data."

When Tunney concluded that day of hearings, he mulled over the best way to build a new forecasting agency—and to ensure that its forecasts would be heeded. Considering Hubbert's experience over the years, this was a conundrum. "His estimates turned out to be accurate with regard to peak production," Tunney said. "Yet the Secretary of Interior, for whom Mr. Hubbert worked indirectly, never knew he existed, never heard of his report."

THE INTERIOR SECRETARY Tunney referred to, Stewart Udall, had since become a friend of Hubbert's, and in 1974 was writing a new book, *The Energy Balloon*. In his book, Udall described how throughout the 1960s he'd heard a "uniformly expansive oil outlook" from both the National Petroleum Council and the Interior Department's experts, including the Geological Survey staff. Udall was disabused of this view only after US oil production peaked and began to decline, and early signs of the energy crisis began to appear. "Having helped lull the American people into a dangerous overconfidence," he admitted, "I felt a moral duty to admit my own errors and to expose the wildly optimistic assumptions that had misled the country."

"In an effort to find out what had happened, I had a long session with Dr. M. King Hubbert," Udall wrote, calling him "a lone pessimist in an industry populated by rabid optimists."

Soon after his testimony in the Senate hearing, Hubbert received another invitation to speak to Congress—this time from Stewart Udall's brother Morris, a member of the US House of Representatives. Morris

Udall, who went by "Mo," had introduced a new bill, the National Energy Conservation Policy Act, which had a central goal of capping the growth of energy consumption at 2 percent a year.

On the first day of hearings on the bill, Udall opened with the observation:

Last year, in 1973, our energy consumption increased by nearly 5 percent over the year before. At this rate we would double energy consumption in fourteen years. This high rate of exponential growth in my judgment cannot be maintained. There is no living system that can tolerate exponential growth indefinitely. And yet our response to the energy crisis has been to say that we need more energy and more growth, the same old approach that got us into the problem in the first place.

Udall continued, noting that some "farsighted scientists, businessmen, economists, and public servants, are beginning to realize that there is a better, safer way than blind, unlimited growth. And that is simply to limit growth now before the problem reaches crisis proportion."

Later that morning, when it was Hubbert's turn to testify, the representative gave him an unusually warm welcome. "I might say that I have a brother who is a big fan of yours," Mo Udall said. "I heard him tell about his years in the Interior Department, saying that he wished he had listened more carefully to you."

Hubbert then launched into a lecture covering his big-picture view, starting from when humanity first developed tools and mastered fire. For millennia, rates of growth—of population, of energy consumption—were very low. In modern times, however, "we have known nothing, particularly in the United States, but exponential growth," Hubbert said. As a result, "we have gone rather naively into the situation we are in," he continued. "It is only within the last ten years that anybody ever questioned this sacred thing of growth." But he saw signs this was changing. "The present proposed legislation is a very important step in our cultural evolution. We have reached the point where we question the desirability of continuing growth."

Hubbert brought all this up to provide a wider context for Udall's goal of capping growth. "Two-percent growth is still an exponential growth with a doubling period of 35 years," Hubbert said. "So even this can only be a temporary transitional phase, but I think a very important one. The most important thing I think we can possibly do is begin to understand this problem. And out of that understanding only then can we hope to achieve rational behavior with regard to it."

"EVENTS ARE NOT proving kind to the USGS position," Philip Abelson had written an editorial in the geology magazine *Eos* in early 1974. Soon afterward the Geological Survey shifted its position, even if only marginally.

In March 1974 the Geological Survey published its latest estimates for future oil discoveries in the United States—and these were lower than their earlier estimates. Whereas Vincent McKelvey's studies, for more than a decade, had held that the nation could produce some 600 billion barrels of oil or more, the survey's latest study generated a range of possibilities for America's ultimate production of 400 to 600 billion barrels.

"The prospects have been judged somewhat less optimistic," McKelvey explained in a press release. Exploration had turned up less oil than expected, he said, adding, "The targets, however, for substantial discoveries are still large."

The survey's latest figures were still more than twice what Hubbert thought reasonable. Nonetheless he saw the new study as a move in the right direction. Hubbert wrote to Wallace Pratt, "For the first time in a decade, the Survey estimates were beginning to be scaled down—not much, but at least the first rung down the ladder."

One oil veteran, though, felt the Geological Survey's estimates were still far too high. In April 1974, John Moody—a petroleum geologist and senior vice president of exploration and producing at Mobil Oil, the nation's third-largest oil company—lambasted the survey in an open letter addressed to McKelvey.

In an unusual move, Moody revealed details of his company's estimates for future oil discoveries. Mobil estimated there were only 13 billion barrels remaining to be found in the contiguous United States. The survey's figures were some ten times larger, with 110 to 220 billion barrels as yet undiscovered. If there really was that much oil to find, Moody wrote, "where in the world (excuse me, the lower 48!) is it?"

For the whole nation—onshore, offshore, and in Alaska—the survey's estimates of undiscovered oil were more than three times higher than Mobil's figures. "This difference is so great that I feel serious attention needs to be paid to it," Moody wrote. If the United States were to base its energy policies on the survey's estimates, he argued, "it would be disastrous for the economy of this country."

To make sure his criticisms were heard, Moody distributed the letter widely, including to Hubbert. "He sent copies of that letter to practically everybody of any consequence in the government," Hubbert recalled. "So there was no burying it under a rug."

John Moody was also a member of a National Academy of Sciences committee known as COMRATE—the Committee on Material Resources and the Environment—which was the successor to the academy's natural resource committees that Hubbert had served on during the 1960s. At the time Moody wrote his open letter, he and a colleague were completing the COMRATE report's section on oil supplies.

Given the clash between Moody and McKelvey, the National Academy committee decided to hold an impromptu one-day workshop to see if it could hash out the reasons for the divergent views. On June 5 the various parties assembled at the academy's headquarters on DC's Mall. Representing the Geological Survey was McKelvey, as well as William Mallory, author of the new study at the center of the debate. Moody was there to represent the oil industry, along with Richard Jodry, a senior scientist from Sun Oil, and a man from the industry-run National Petroleum Council. Hubbert attended, too—but as usual, whenever he spoke about energy supplies, he represented only himself.

Mallory opened the workshop by explaining his approach. As with most of the Geological Survey's estimates over the past decade, Mallory had used a variation on Zapp's old method. He argued Zapp's assumption—that future drilling would be as successful as drilling in the past—was "optimistic but not unreasonable." He also included a lower estimate for a "conservative" case, in which future drilling turned out to be only half as successful as the historical average. Anything less than that, Mallory ruled out, arguing it was "pessimistic."

"Well, that's one hypothesis," Hubbert thought. According to his 1967 study, "Degree of Advancement of Petroleum Exploration," the rate of discovery had already dropped to under a one-third of the historical average, clearly in Mallory's "pessimistic" range.

Moody presented Mobil's results, which were based on the geology of the remaining unexplored areas, and comparisons with analogous areas that had already been explored. Mobil had used a completely different method than Hubbert's, yet the results were remarkably similar. Jodry of Sun Oil chimed in that he'd also made estimates of undiscovered oil, coming up with numbers in line with Mobil's and Hubbert's. These industry representatives both argued that the Geological Survey's approach was overly optimistic, because the success rate of future drilling would likely be far lower than the historical average. Though they didn't cite Hubbert's 1967 study, the pattern these industry insiders described was exactly what he had found. The Zapp hypothesis, they agreed, was wrong.

The National Petroleum Council had come up with high estimates of oil, close to those of the Geological Survey. But at this meeting, the council's representative pointed out that its latest estimate had been based on an assumption that the industry would eventually recover 60 percent of the oil in the nation's fields. If instead the recovery were only one-third—the historical average—then the council's estimates were in line with Moody's, Jodry's, and Hubbert's.

Despite these challenges from the industry, McKelvey and others at the Geological Survey stuck by their results. When a reporter asked what he thought of Hubbert's criticism, McKelvey said he was "mulling it over."

Blueprints

THE UNITED STATES COULD ACHIEVE energy independence—or at least, reduce its dependence on foreign oil to a tolerable level—according to a massive report called the *Project Independence Blueprint*. Issued in November 1974, this report was the central task of a new agency, the Federal Energy Administration, the successor to the Federal Energy Office that President Nixon had created during the OPEC embargo. Before the *Blueprint* was complete, though, Nixon resigned from the White House, under the cloud of the Watergate scandal. His successor, Vice President Gerald Ford, maintained the goal of energy independence—although with the deadline pushed out from 1980 to 1985.

It was possible for the United States to cut its oil imports to almost nothing, the report argued, because the nation's oil production could achieve a revival, soaring to all-time highs. This outlook was based on forecasts generated by a task force led by the Geological Survey's director, Vincent McKelvey. According to these forecasts, if oil prices remained as high as they were, at $11 per barrel, US oil production could increase 50 percent by 1985, reaching a new all-time high at 15 million barrels a day. If the nation pursued "accelerated development," with essentially all sources of oil pushed to the maximum, then by 1985 production could soar even higher, to 20 million barrels a day—and on top of that, the

United States could produce another 1.5 million barrels per day from oil shale and from turning coal into liquid fuel.

Buried in the middle of the report, however, was a warning that "this level of production could not be maintained indefinitely at these prices, as oil reserves at these prices would soon peak." At the end of the report, a section titled "Beyond the Year 1985" looked out more than a decade. Across a range of scenarios, "the oil and gas shortfall is so large," it said, the nation would need to curtail its use of oil and gas, using energy more efficiently and also relying more on electricity. The report called for long-term research and development on "new energy sources not limited by conventional fossil fuel and uranium resources," such as nuclear breeder reactors, nuclear fusion, and solar power, which would "take decades to develop and introduce." Boosting oil production might allow the United States to achieve energy independence for the short run but wouldn't be a long-term solution.

UNDER PRESIDENT FORD, Henry Kissinger and James Schlesinger retained their positions as secretaries of state and of defense. They hadn't followed through on their plan to invade Abu Dhabi. But with oil prices remaining persistently high, and US crude oil imports soaring, the notion of invading an oil-exporting nation still held appeal.

"One of the things we hear from businessmen is that in the long run the only answer to the oil cartel is some sort of military action," *Business Week* asked Kissinger in January 1975. "Have you considered military action on oil?"

"A very dangerous course," Kissinger replied. "I am not saying that there's no circumstance where we would use force. But it is one thing to use it in the case of a dispute over price, but it's another where there is some actual strangulation of the industrialized world."

This comment sparked debate and speculation around the world, with other nations calling for more explanation. (One headline read: "Actual Strangulation Requires Definition.") But the White House declined to comment. Finally two weeks later, in an interview with *Time*, Ford said, "I would re-affirm my support of that position." When asked what would count as "strangulation," Ford said it was when "you are just about on your back."

Soon after, in March 1975, an article appeared in *Harper's* magazine titled "Seizing Arab Oil," laying out a plan for the United States to invade the Middle East's largest oil producer, Saudi Arabia—and arguing that it must do so. After taking control of Saudi oil fields, the United States could boost production and drive oil prices down, to the benefit of Americans and consumers around the world, the article claimed. The invasion would require a huge force—a division of Marines, 14,000 troops. For the subsequent occupation, the forces would ramp up to 40,000 troops, backed by air support to fend off fighter jets from Saudi Arabia and its neighbors, and a continued military presence to prevent sabotage by locals. With lower oil prices worldwide, the money America saved on oil imports would more than pay for the operation, the article argued. In any case, the developed nations had no choice, since "no one denies that the dependence of the Western world on Arab oil is absolute."

The article was credited to "a Washington-based professor and defense consultant with intimate links to high-level US policymakers," using the pseudonym Miles Ignotus—Latin for "unknown solider."

Following on such ideas, the Congressional Research Service published a 106-page report, *Oil Fields as Military Objectives: A Feasibility Study*, which included maps of the oil fields across the Middle East and North Africa, as well as Venezuela (another OPEC member and a major supplier to the United States). While OPEC members' militaries "could be swiftly crushed," the report concluded, occupying their oil fields would be far more difficult, requiring "constant security against skulduggery," so the whole operation would "combine high costs with high risks."

James Akins—by then the US ambassador to Saudi Arabia—challenged such talk of an invasion, saying, "Anyone who would propose that is either a madman, a criminal, or an agent of the Soviet Union."

Only afterward, Akins said, did he find out that the article was based on briefings by Kissinger—his boss. Later that year Akins was fired.

. . .

IN THIS MIDST of this discussion of invasions, the National Academy of Sciences released its COMRATE report. Its section on oil supplies, led by Mobil's John Moody, strongly backed the lower end of the range of estimates. The leader of the whole report, a Yale geology professor, called for the nation to adopt energy conservation as "kind of a national religion." *Business Week* declared the COMRATE report "clearly draws its inspiration from M. King Hubbert" and detailed the ongoing fight between Hubbert and McKelvey. "Hubbert's method is tied to the past," McKelvey said in defense of the survey's work. "I don't think you can figure out what is in the ground on that basis."

The Senate and House then held a joint hearing titled "Adequacy of US Oil and Gas Reserves," which questioned the Geological Survey's work, contrasting it with the COMRATE assessment. In the congressional hearing, McKelvey said there had never really been any official Geological Survey estimates—only studies done by individual researchers within the survey, who might not agree with each other. He also applied caveats to the survey's estimates, saying they were "in reality are targets for exploration, rather than projections of the amounts of oil and gas that will ultimately be produced."

A Commerce Department panel also criticized the Project Independence report, saying it gave "the impression that it will be easy to achieve energy independence." The panel concluded "there is a danger of being lulled into a sense of complacency."

As Hubbert recalled, "The realization had spread through the government that the Geological Survey had been misinforming them for the last decade or so, and McKelvey was the most visible, overt person involved in the thing." So, Hubbert said, around this time McKelvey was getting "very, very jumpy."

THIS CONTROVERSY GAVE the Geological Survey's Pick and Hammer Club a much juicier target than usual for its annual satirical skits, which regularly skewered the survey's researchers and administrators. In the April 1975 show, one skit featured "Vincible McWelldry, The Top Gasser" and "Em King Blubbert," called the "oracle" of groundwater and "all moving liquids." As McWelldry and Blubbert engaged in a long-winded debate, roustabouts brought a wind turbine onto the stage, and a woman announced, "Gentlemen, if you'll just continue your argument indefinitely and face the fan, there'll be no more energy crisis."

The woman in the Pick and Hammer Club skit probably represented Betty Miller, one of the few female geologists at the survey, as well as the lead author on a report then undergoing harsh internal criticism.

Prior to the OPEC embargo, but as signs of an energy crisis were piling up, the Geological Survey had begun trying to improve its assessments of oil. It brought in new staff, including Peter Rose, a geologist from Shell Oil, hired to head up the survey's Branch of Oil and Gas Resources. On starting there, Rose realized "they didn't know a goddamn thing about oil and gas," he recalled. "The Survey had gradually allowed its expertise in oil and gas exploration and oil and gas resources to atrophy. It was just criminal, what they had allowed to happen. My coming on board was just a frantic attempt to begin to shore up about twenty-five years of poor management."

As the Federal Energy Administration was wrapping up its *Project Independence Blueprint*, it commissioned the survey to take a fresh look at America's oil resources. The survey got a Shell veteran, Harry Thomsen, to come out of retirement to oversee that study, and Betty Miller, who had fourteen years of petroleum industry experience prior to joining the survey, was lead author.

Rather than using a version of the Zapp approach, which the survey had employed in its national oil assessments for more than a decade, Miller and Thomsen started from scratch. They evaluated 102 individual provinces, drawing on the survey's experts' knowledge about the distinct geology of each province. The team then assembled a range of estimates of oil in each region, rated in terms of the likelihood they would turn out to be correct. Summing up the results for the whole nation, including Alaska, they were very confident—19 chances out of 20, they figured—

that the United States would ultimately yield at least 218 billion barrels of oil. There was a fifty-fifty chance that the nation would ultimately yield 250 billion barrels. And there was a small chance, only 1 in 20, that the ultimate production would be more than 295 billion barrels.

The study didn't directly critique either the Zapp hypothesis or McKelvey's work based on that approach, but it did offer far smaller numbers than the survey's previous studies. The new results implied there was almost zero chance that the survey's earlier estimates—including McKelvey's own—would turn out to be correct.

As with all survey reports, Miller's had to get approval from the director, McKelvey. When she turned in her report, she recalled, "I got an irate call from McKelvey and one of his henchmen. I was accused of joining 'the enemy camp.'" She thought, "Enemy camp? We're straight scientists."

"The Director denied the results, and pressure was put upon me about changing the results, which I absolutely refused to do," Miller recalled. "They threatened my job." When she replied that she'd happily return to industry, they backed off. McKelvey then wanted to bury the report. But the Federal Energy Administration, which commissioned the report, had already seen drafts of it. "They made no bones about it to the Survey, that if you don't let us have those results that we've paid for, we're going to blow it sky-high," Miller said. Plus there was Hubbert. As Miller recalled, "They knew Hubbert was waiting in the wings to blow the whistle."

Miller hadn't known of Hubbert's work while compiling the report. But soon after finishing it, she came from Denver, where she was based, to the survey's headquarters in DC, and on arriving Hubbert invited her to come to his office. He laid out the history of his battles with McKelvey and showed Miller the report he'd written for the Senate, which harshly criticized McKelvey. "How in the world did you get approval to have this published?" Miller asked. He looked her in the eye and smiled, saying, "I report to a higher authority."

Miller realized her estimates were in the ballpark of Hubbert's—and that he saw Miller's study as vindication of what he'd been saying for decades, despite Miller having used a much different approach than his own. "He appreciated and approved the results of our studies, but he didn't care for our methods," Miller recalled with a laugh. "That's all right. We didn't care for his methods."

After learning of this long history, Miller finally understood McKelvey's opposition to her own work. From then on, she felt that McKelvey and his allies were driven not by the science but by "professional jealousies and vindictiveness."

With little choice but to release Miller's report, McKelvey went over the authors' heads and tampered with the results, the authors told *National Journal.* For offshore areas in Alaska and along the Atlantic coast, the study had estimated that the recoverable oil might turn out to be as low as zero barrels—that is, no recoverable oil in either region. In the final report, these numbers for those regions were changed to a tighter range of uncertainty than used in the rest of the report, so that the low end of the range for Alaska's offshore oil could be raised to 1 billion barrels, and to 2 billion barrels for the Atlantic offshore. The authors' take was that McKelvey and others had changed these numbers because "a zero recovery rate isn't very promising for such strategic areas."

Despite being watered down slightly, when Miller's report went public, it made nationwide news with headlines like "Oil, Gas Estimates Slashed in Survey." But when McKelvey spoke publicly about the study, he tried to spin the results, arguing the new, lower estimates would not change the forecasts his task force had created for Project Independence. "Even if the low range of the estimate should prove correct," McKelvey told a reporter, "its dimension would not be limiting until near the end of the century."

McKelvey seemed to be saying the nation could produce plenty of oil through the year 2000—yet US production had already been in decline for several years.

HUBBERT FELT THAT Miller and Thomsen had done "a heroic job," persevering through a "bloody battle" to get the report released. The new study was "the most important report issued by the US Geological Survey during the last fifteen years," Hubbert told another oil industry veteran at the time. (The survey had never published any work by Hubbert during the dozen years he'd been on staff there.)

After the Miller-Thomsen report appeared, *The New York Times* came to Hubbert's house to interview him—and he talked for three hours. The newspaper of record ran a profile of him that opened, "When the United States Geological Survey acknowledged that it had been overestimating the nation's oil and natural gas resources, M. King Hubbert might have allowed himself a discreet smile of satisfaction."

The *Times* cited respected figures—Philip Abelson of *Science* and the political powerhouse of Stewart and Mo Udall—who praised Hubbert for having nailed the timing of the peak and decline of US oil, and for trying to warn the nation. Hubbert, the paper said, had achieved a status as "a prophet without sufficient honor." *Science* covered the controversy over oil estimates in its own news article, saying Hubbert had "acquired a reputation as something of an oracle."

In his role as oracle, Hubbert received increasing requests to deliver talks or provide his judgment on energy-related matters. A CIA analyst called to ask about the North Sea, a major offshore province shared by the United Kingdom, Norway, and Denmark, where production was just starting. The analyst wondered if the early estimates of oil there— some 20 or 30 billion barrels—might undergo the kind of reserve growth that Hubbert had parsed in his 1967 "Degree of Advancement" paper. There, he'd calculated that fields would undergo, on average, about sixfold growth in their reserves, from initial estimates to final tally.

This sixfold rule of thumb—based on US fields discovered in the past, mostly onshore—wouldn't apply to the North Sea, Hubbert figured. "With modern exploration technology," he told the CIA analyst, "seismic crews can cover the North Sea like plowing a corn field." Companies could quickly compile a comprehensive map of the undersea geology, so they were likely to get a much better picture from the start of how much oil was there. He thought the North Sea's eventual reserve growth would be much smaller—perhaps a factor of two rather than six. Hubbert's approach for estimating future reserve growth was powerful—but if used sloppily, it could easily generate misleading results.

Fielding many requests to speak, Hubbert delivered dozens of talks to diverse groups—from the Asphalt Manufacturers Association to the American Butter Institute. The US Department of Health, Education, and Welfare taped a long interview with Hubbert, which it distributed to hospitals. Showing an idealized curve for world oil, Hubbert explained, "The OPEC countries are tampering with this curve right now—they're actually curtailing production somewhat." As a result, world oil production would likely peak at a lower rate and at a later date. OPEC's curtailment would stretch out the whole curve "by seven, eight, ten years, maybe. But it doesn't alter the basic thing that I'm saying significantly." Still urging a transition off fossil fuels, he brought along a small gadget that fit in the palm of his hand: a tiny solar panel that drove a fan, demonstrating solar energy in action.

BY 1976, HUBBERT had hit the limit of fifteen years of service for a federal employee in a civil-grade appointment. To figure out exactly when he had to retire from the Geological Survey, someone tallied up all his earlier work for the government, going back to his work as post office clerk in Chicago in the 1920s, and human resources informed him that he'd have to retire on August 31—two weeks earlier than he'd planned. "What the hell is going on now?" he wondered. Somebody, he suspected, had ordered human resources to go back through the old records, as he couldn't imagine someone taking it upon themselves to do that.

As the time of his retirement approached, Hubbert made plans for saying goodbye. Usually longtime researchers had a retirement dinner party, but Hubbert thought that would be "rather embarrassing," he wrote to Thomas Nolan, who'd recently retired from the survey himself. Instead Hubbert went out to lunch with a few friends. He also held an "open house" one afternoon, where colleagues could stop by to chat and have drinks. Someone drew up a simple flyer for the event, reading "Goodbye King," with a sketch of a man holding a pointer aimed at the crest of a large, rounded curve with a crown perched on top, labeled "Hubbert's Pimple."

Buoyed Up

IN JANUARY 1977, JIMMY CARTER—the new US president, having defeated Gerald Ford—set an ambitious goal for his administration: creating a new national energy policy within ninety days of taking office.

Before the White House released a description of the plan, reporters uncovered some of the details—as well as the underlying motivations. The "central premise" was that "the world's supply of oil is finite, it is running out, and no easy substitute is in sight," as the influential *New York Times* columnist Anthony Lewis put it. These notions were nothing new, Lewis noted, and the shortages America then faced had been forecast years in advance—in particular by Hubbert. "But it is very new," Lewis added, "for the United States Government to accept Dr. Hubbert's thesis as the premise of its policy."

Carter had become deeply worried about the future of oil supplies. His main influence in this area was Admiral Hyman Rickover, who had been Carter's commanding officer when he served in the navy. In his 1959 book *Education and Freedom*, Rickover had warned of an approaching peak of US oil production around 1965—the same as Hubbert's favored forecast at the time. Overall, Rickover's general views on energy and its central role in civilization were much like Hubbert's.

As his go-to man on energy, Carter chose James Schlesinger, the energetic and brilliant economist who had served most recently as secretary of defense, before getting fired by Ford. Earlier Schlesinger had held a wide variety of roles: director of the Bureau of the Budget, chairman of the Atomic Energy Commission, and CIA director. Back in the White House under Carter, Schlesinger had the mammoth task of overseeing the creation of the nation's new energy policy—an assignment *Time* magazine dubbed the "Superbrain's Superproblem." Then in his late forties, thick-necked and prematurely white-haired, and incessantly puffing on a pipe, Schlesinger had to try to convince Congress to approve the Carter administration's plans.

One major step would be to consolidate the nation's hodgepodge energy agencies. There was the Bureau of Mines that monitored coal and other minerals, the Federal Power Commission that oversaw electricity markets, the Federal Energy Administration that had grown from efforts to cope with the 1973 oil embargo, and the Energy Research and Development Administration, a reformulated version of the Atomic Energy Commission. Carter proposed combining them all in a new Department of Energy, and put forward Schlesinger as his candidate to join the cabinet as the nation's first energy secretary.

Since the end of the OPEC embargo in early 1974, oil prices had remained persistently high, but Americans seemed to have grown used to these high prices—even as the economy continued to sputter. So Schlesinger and Carter had to convince the nation that its energy problems had not been solved, that they had only temporarily disappeared from the headlines. As *Time* put it, there was "as yet no national consensus that a real crisis exists."

"Americans have never understood finiteness," Schlesinger declared at the time. "They have believed in growth, expansion, limitless resources. In energy, all of those presuppositions must perforce change." Nonetheless, he was hopeful that when faced with the facts, the nation could have a rational discussion and come up with solutions. "The basic question that faces us on energy," he said, "is not the particular techniques but whether we are serious. And the country has to be serious, because otherwise Washington will not be."

. . .

WHILE THE CARTER administration was preparing its new policy—to be announced in a major speech—Schlesinger's office phoned Hubbert's house seeking permission to quote from his work in Carter's speech.

Hubbert was long gone. He'd left a couple of months earlier, on a four-month lecture tour across Europe, hopping from embassy to embassy, speaking to diplomats, high-level officials, and top academics. It had been arranged months earlier under the Ford administration, by the US Information Agency, which ran America's public diplomacy throughout the Cold War.

On the eve of Carter's energy speech, a reporter at *Newsday* did manage to reach Hubbert on the phone in Copenhagen, Denmark. About being sent on tour by the US government, to discuss his forecasts and warnings that the same government had ignored for years, Hubbert told the reporter, "It is a little ironic, isn't it?"

BEFORE CARTER PROPOSED his new policies to Congress, on April 18, 1977, he presented them to the American people.

"Tonight," Carter opened, "I want to have an unpleasant talk with you about a problem unprecedented in our history. With the exception of preventing war, this is the greatest challenge our country will face during our lifetimes. The energy crisis has not yet overwhelmed us, but it will if we do not act quickly."

When it came to oil supplies, there was no quick fix, he argued, and the problems would persist—and likely grow worse. "In spite of increased effort, domestic production has been dropping steadily," he explained. Later in the speech, he reiterated, "We can't substantially increase our domestic production."

Before long, the whole world would face a similar situation. "Each new inventory of world oil reserves has been more disturbing than the last," Carter said. He then ventured a forecast: "World oil production can probably keep going up for another six or eight years. But some time in the 1980s it can't go up much more. Demand will overtake production. We have no choice about that."

But the country did have a choice about how to deal with the situation, Carter declared. America could "drift along," continuing to be "the most wasteful nation on earth." Or it could take action. To cope with dwindling production at home and prepare for this worldwide crunch, Carter called for a suite of new policies. The nation could rely more heavily on coal, its most abundant energy source, and also develop "permanent renewable energy sources like solar power." But even if the nation raised energy production, it would also have to conserve. "Conservation is the quickest, cheapest, most practical source of energy," he said, and set a goal of insulating nine out of ten American homes and all new buildings.

Carter also called for removing controls that held down prices of oil and natural gas. "Prices should generally reflect the true replacement cost of energy," he argued. "We are only cheating ourselves if we make energy artificially cheap and use more than we can really afford."

Although improving efficiency could save consumers money, overall these efforts would require sacrifice and some change to American lifestyles. "The alternative may be a national catastrophe," Carter argued. "This difficult effort will be the moral equivalent of war."

OVER IN EUROPE, Hubbert didn't catch Carter's televised speech. But he read a transcript from the US embassy in Stockholm, underlining passages and scribbling comments in the margins. Reading the speech, Hubbert stopped at one part in particular: "During the 1950's, people used twice as much oil as during the 1940's. During the 1960's, we used twice as much as during the 1950's. And in each of those decades, more oil was consumed than in all of mankind's previous history."

In the margin next to that bit, Hubbert wrote his initials, "MKH." It wasn't a direct quote, but it seemed to be pulled from one of his articles.

"I'm excited over Jimmy Carter's speeches," Hubbert wrote to his old mentor, Harlen Bretz, from Paris. "It is the first time that the federal government, at the top, has ever faced up to the real situation we are in."

But at the same time, Hubbert found parts of the speech frustrating. One of the main principles in Carter's plan was that "healthy economic growth must continue." Hubbert underlined that bit, writing in the margin "Growth!"

Also, Carter's gloomy forecast for world oil—that its production would peak in the mid-1980s—didn't fit with Hubbert's outlook. He expected the peak to come closer to the year 2000.

Carter's world oil forecast was based on a CIA assessment. But Schlesinger, having served as CIA director, didn't trust the agency's analysis and told Carter, "Don't let that thing become public knowledge!" Carter had overruled him—and took the unusual step of declassifying the report's summary and releasing it at the time of his speech. "The CIA report is another reminder that world oil production limits will be reached in the 1980s," said a White House press release. A major reason for this gloomy view was that "the Soviet oil industry is in trouble," the CIA concluded. "Soviet oil production will soon peak, possibly as soon as next year and certainly not later than the early 1980s."

Hubbert also got a copy of that CIA report from the Stockholm embassy. Next to its section on Soviet oil, he wrote, "Sweeping statement with no supporting evidence whatever."

Hubbert's feelings of vindication outweighed such frustrations. The day after Carter's speech, Hubbert gave a lecture in Stockholm. The local US embassy reported in a cable to DC that he had given an "exciting, rapid-fire summary" of world energy. "Seminar attended by overflow audience of 90 experts. List reads like Who's Who in Swedish energy community. Audience fascinated, impressed by Hubbert's knowledge and ability to view energy perspective in vast perspective."

In the days that followed, Hubbert moved on to Brussels. He was "obviously buoyed up" by Carter's speech, the Brussels embassy cabled home, and was no doomsayer. "Hubbert is optimistic about American people's growing awareness of the gravity of energy problems and willingness to make sacrifices now that they are being given leadership."

Incomprehensibly Large Numbers

EACH TIME A NEW PRESIDENT took the White House, it was customary for the heads of departments and agencies to tender their resignations. For agencies seen as nonpolitical, such as the Geological Survey, these resignations were usually turned down and the directors retained. But Carter left McKelvey in limbo, his resignation neither accepted nor rejected.

In the spring of 1977, when Hubbert returned from his European lecture tour, he heard that the National Academy of Sciences, which usually made recommendations for new Geological Survey directors, had been asked to search for a replacement for McKelvey. Nothing was official as yet, but it looked like McKelvey was on his way out.

Such talk remained only rumors until July, when McKelvey's superiors in the Interior Department informed him he'd been dismissed. As McKelvey later relayed it to journalists, the new interior secretary "wanted his own team." Officially McKelvey had resigned, but in reality he'd been fired—the first director to be fired in the Geological Survey's century-long history.

Days after McKelvey got word—but before the news had gone public— he gave a speech in Boston to a group of investors in which he presented his most expansive estimates yet. He talked of vast quantities of natural gas in tight sandstone in the Rocky Mountains, of gas in black shales along the eastern seaboard, and gas trapped in coal beds across the country. The greatest prize, he argued, was a resource known as geopressured methane in the Gulf of Mexico, which held an estimated 60,000 to 80,000 trillion cubic feet of gas. "This is an almost incomprehensibly large number," McKelvey said. "Even the bottom range represents about 10 times the energy value of all oil, gas and coal reserves in the United States."

What McKelvey didn't highlight, though, was that no one had yet tapped such geopressured methane in any significant amounts. The estimates were highly uncertain, a bold extrapolation from limited data.

WHEN HUBBERT HEARD that McKelvey had been fired, the reason was clear to him: McKelvey's petroleum estimates were biased, and in his optimism he'd encouraged the nation to walk blindly off a cliff. Hubbert wrote to a friend that McKelvey's tenure had been "the most disgraceful epoch" in the survey's history.

However, others in the scientific community disagreed. While sympathetic to Hubbert's warnings, *Science* editor Philip Abelson argued McKelvey's dismissal was a dangerous sign of politicization. Thomas Nolan, the former Geological Survey director, voiced the same worry. A *Washington Post* column likewise fretted, "It would be deplorable if that much-depended-upon agency were to be pushed into fudging resource estimates."

Hubbert heard similar notions from others in the survey, but he didn't buy it. Writing to Harlen Bretz, he said:

Apparently the story being put out by McKelvey and his inside clique at the Survey is that this took them all by surprise, and that they just can't imagine why this happened; it must have been political. The fact is that McKelvey has barely been hanging on for the last 4 years because he got caught in the misinformation he had been feeding the government since 1961 about the supplies of oil and gas in the United States. We have all been expecting this news ever since the Carter Administration took over. McKelvey has known this as well as everyone else, so his present attempt to give it a political flavor is simply a continuation of the deceptions he has been engaged in for the last 16 years.

In September, when the Interior Department publicly announced McKelvey's "resignation," some observers agreed with Hubbert that the reason was McKelvey's persistent optimism. But almost everyone who made this argument thought McKelvey had been correct all along. One of the first defenders was an *Oil and Gas Journal* columnist, who lauded McKelvey for having "repeatedly expressed his strong views that there is plenty of petroleum still to be discovered in this country and the search should be pressed aggressively."

Politicians also came to McKelvey's aid. An early supporter was Jack Kemp, a former football star. As a Republican congressman representing New York, Kemp had led a vigorous fight against Carter's energy plans. After McKelvey's firing, Kemp charged in a speech in the House that it had been "because of his optimistic view of our energy situation—a view that runs counter to the Malthusian pessimism of President Carter."

McKelvey also received support from another public figure: actor-turned-politician Ronald Reagan, until recently governor of California. In 1976 Reagan had come close to beating Gerald Ford for the Republican nomination for president. While biding his time for another run at the White House, Reagan had a daily radio show, syndicated nationwide, in which he delivered a short speech. Reagan devoted one show to defending McKelvey, presenting a parade of numbers that suggested world oil estimates, over the years, had simply gone up and up. Reagan cited a 1942 estimate of 600 billion barrels, a 1970 estimate of 4 trillion barrels, and as a finale added "Russian scientists now place it at 12 trillion." His conclusion: "We don't need a new energy department, we only need deregulation."

The Wall Street Journal took up the cause with a long editorial, "Good Bye, Dr. McKelvey." "While you are trying to bulldoze a $100 billion tax energy conservation program through Congress," the editorial said, "it does not help to have the government's top expert on such matters running around talking about the vast resource base in the United States."

This defense of McKelvey was in line with a string of editorials *The Wall Street Journal* had been running since the start of the Carter administration. These pieces repeatedly equated forecasts for a production peak with expectations that energy would suddenly "run out" and that "civilization will grind to a halt," as one editorial put it. "This is nonsense," the *Journal* concluded. (No one credible—not Carter, not Schlesinger, not Hubbert—had actually made such claims.)

The *Journal* claimed there were enormous resources waiting to be tapped. An editorial titled "1,001 Years of Natural Gas" argued, just as McKelvey had, that geopressured methane could provide the United States with plentiful supplies for centuries. Another editorial said Russian scientists had estimated there were vast amounts of natural gas on the planet—including quantities dissolved in oceans and swamps—"20 million years worth, conservatively." They concluded: "There are plenty of hydrocarbons to last until solar power comes in, so long as we will pay the cost of retrieving them." Their solution was simple: "All that's needed is to deregulate prices, getting the clumsy government out of the fray."

The complacency the *Journal* expressed about resources was shared by many Americans. After Carter's "moral equivalent of war" speech, Gallup polls found the percentage of Americans who agreed there was an energy crisis increased from 43 to 54 percent—but that implied about half of them either thought there was no energy crisis or didn't know what to think about it. Later that year another poll revealed that roughly half of Americans understood that the United States imported any oil at all.

The Prophet

"IT WAS NOT UNTIL AFTER the 1973 oil crisis that Hubbert's scientific achievement was recognized," said the official announcement of the Rockefeller Public Service Award, in naming Hubbert one of its winners for 1977. This award, celebrating work for the federal government, went to Hubbert for his forecasts, which had "laid the foundation for US energy policy."

That may have been an exaggeration, since Carter was influenced heavily by Hyman Rickover's ideas rather than Hubbert's. But it was true that the Carter administration, far more than any before, had based its energy policies on an outlook like Hubbert's. James Schlesinger, by then confirmed as the United States' first energy secretary, was well aware of Hubbert's forecasts for peak oil production. And there were those Hubbert-esque lines in Carter's speech about the doubling of energy consumption.

In November, when the Rockefeller Public Service Awards were publicly announced, *The Washington Post* ran a front-page profile of Hubbert, "Oil Prophet Cited: Geologist Saw Crisis in '48." The piece was reprinted in many other newspapers, and as a result, Hubbert told an old friend, "all hell has been busting loose and I am swamped with mail." More journalists wanted to write about him, and an NBC producer took him to lunch to talk about having him on the *Today* show.

Throughout the 1970s, Hubbert had been sporadically called a "prophet" and an "oracle." After the award and the *Washington Post* profile, such appellations became commonplace. Replying to friends who sent letters of congratulation, Hubbert wrote, "So accustomed am I to playing the role of 'minority of one'"—a title Philip Abelson had bestowed on Hubbert years earlier—"that I have not yet found words appropriate to the opposite role of 'acclaimed prophet.'"

WHEN CONGRESS PASSED the Department of Energy Act in October 1977, integrating the nation's various energy agencies under one umbrella, it also established a new body called the Energy Information Administration (EIA). Although part of the Department of Energy, the EIA was intended to be independent from other research or priorities within the government and was assigned the broad task of studying "the adequacy of energy resources to meet demands in the near and longer term future."

However, John Dingell, a Michigan Democrat in the US House, felt the government didn't have the capacity to do systematic and objective analyses with a truly long-term outlook. He thought existing nonprofits weren't up to the task either. So in April 1978 Dingell introduced a bill—cosponsored by a young Tennessee Democrat, Al Gore—to establish and fund a National Energy Policy Institute, a nonprofit that would provide the government with energy advice. As Dingell put it, this new institution would be "charged with looking, not just ahead, but over the horizon."

Dingell invited Hubbert to testify about the bill. In retirement, Hubbert no longer had to ask for permission to speak, so he appeared on the first day of the hearings as the initial witness. Dingell opened the hearing by saying:

> In the past, we acted as though we could afford to ignore the distant future, leaving subsequent generations to attend to their own problems. The Nation has muddled along, lurching from crisis to crisis, and avoiding catastrophe only because our world had enough lati-

tude and our systems had enough flexibility to adjust for the effects of incompetency and profligacy. In short, we acted as though we lived at the open end of a natural horn of plenty.

It is now clear to all, even the most dewy-eyed optimist, that we can no longer afford such luxury. The basic energy resources upon which the Nation has relied have now clearly been shown to be finite.

When it was Hubbert's turn to speak, he began with praise. "As I read this bill," Hubbert said, "it impressed me as being potentially one of the most important proposed pieces of legislation that have come to my attention recently." Although a handful of universities had been working on long-term analysis, they were only just beginning this task, he said. "Hence, in my opinion the institution proposed by this bill is urgently needed."

However, Hubbert argued, "the success or failure of this institution depends to a very large extent upon the quality of the director and the initial technical staff." It wasn't enough to simply recruit experts with "impeccable academic credentials." If the institute employed experts who saw industrial society in physical terms, recognizing limits to fossil fuels, and who hailed from many disciplines—such as anthropology and sociology—they could think creatively about the long-term outlook. In this case the new agency could become "a very powerful institution," he argued. "They are going to be in very heavy demand by the Government and by the public to keep them informed on what is going on, and what is the outlook for the future."

In discussing the proposed energy policy institute, Hubbert said, "We have institutions that are supposed to be dealing with problems of this kind. To name one, Resources for the Future had an original assignment of this nature." That organization had sprung from a major study commissioned by President Truman and published in 1952 under the title "Resources for Freedom." The report had considered the possibility that US oil production might peak in the 1960s and was generally cautious about the future availability of resources. Yet the organization Resources for the Future had gone on to issue some of the most expansive estimates and highest forecasts on record. Their main problem, Hubbert argued, was an adherence to a dogmatic economic approach that did not recognize limits to resources or to growth.

Here the head of the subcommittee's staff, Frank Potter, spoke up. "Resources for the Future have been called by some of its less sympathetic observers as 'Resources for the Very Immediate Future,'" Potter quipped. "It is not my place to knock economists, because, God knows, I do not understand them," he added. "I am sure they serve a very useful function someplace." He agreed with Hubbert that the proposed institute would need experts who could escape from the usual economic approach. "I think you have really put your finger on the core problem," he told Hubbert. America had become so accustomed to exponential growth that it had become a part of everyday life, Potter added. "It is very difficult to get outside of that."

Hubbert had been making that argument for nearly fifty years.

Hubbert Factors

IN THE LATE 1970S, THE relatively new Energy Information Administration was developing new methods for its forecasts for oil in the United States and worldwide. From the start, this arm of the Department of Energy, tasked with gathering statistics and looking into the future, followed Hubbert's lead in crucial aspects of its analysis.

For gauging the prospects for US oil production, the EIA tried employing Hubbert's early approach, computing a bell-shaped curve that fit the history of production so far. As applied by the EIA, it gave the result that the United States would ultimately yield around 160 billion barrels of oil—lower than Hubbert's estimates. (However, the report did include caveats: technological surprises or rising prices could lead to higher production.) Also, when the EIA adjusted its oil estimates to take account of future reserve growth, it used what it called "Hubbert factors."

At a National Bureau of Standards workshop on energy forecasts, a Princeton statistician attacked Hubbert's work as "extremely weak" and argued that his "geological, economic and technical assumptions go unsupported and unchallenged." One of the EIA's top oil analysts, Charles Everett, explained he'd invited Hubbert, who couldn't make it. So in Hubbert's defense, Everett made a simple rebuttal: "One fact: Domestic crude oil production peaked in 1970, so Hubbert was very close."

Looking at world oil prospects, the EIA also cited Hubbert's work and used a graph almost identical to one he regularly presented. It showed two possible scenarios for world oil: either a sharp peak and decline, or a long plateau for decades and then a decline. The EIA expected that "actual production will probably fall between these two extremes." Its outlook assumed that the world would ultimately extract 2,100 billion barrels of petroleum liquids—both conventional oil and natural gas liquids.

Back in 1948, in his talk before the American Association for the Advancement of Science, Hubbert had used an estimate of 2,000 billion barrels for the world's ultimate production of oil. Through the 1950s and 1960s, Hubbert had become more pessimistic, using numbers as low as 1,250 billion barrels. But then as more evidence came in, he had raised his sights again. By the 1970s he had returned to that earlier figure of 2,000 billion barrels. Subtracting natural gas liquids, the EIA's

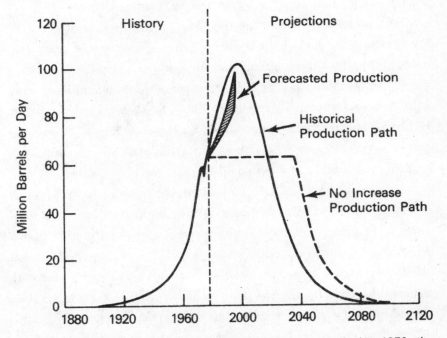

World oil production forecasts: Building on Hubbert's ideas, in the late 1970s the EIA argued that world oil production would likely peak around 2010.

figure was equivalent to about 1,800 billion barrels of crude oil, so the EIA was somewhat more pessimistic than Hubbert.

If such an estimate was correct, the EIA report said, "total world oil production would start to fall at some point during the 1995–2035 time period." Given that the world did appear to be following a path between the two extremes, the EIA argued, this suggested world oil production would peak around 2010.

SOON AFTER THIS forecast was issued, such predictions had a monkeywrench stuck in their gears.

"Iran is not in a revolutionary or even a 'pre-revolutionary' situation," the CIA had reported in the summer of 1978. The Defense Intelligence Agency concurred that Iran's king, or shah, Mohammad Reza Pahlavi, was secure. Iran was a central player in world oil markets as the non-Communist world's second-largest oil exporter, and the shah was a key US ally. So Iran was closely watched—but not closely enough.

By the end of 1978, an uprising had gathered strength, and early the next year the shah fled the country, ending his reign of nearly four decades. The Iranian revolution blindsided America, regular citizens and the government alike. Iran's oil exports plummeted from 5 million barrels a day to zero. Other major producers—including Saudi Arabia, Kuwait, and Iraq—boosted their production, but it wasn't enough to offset Iran's decline. With supplies falling short of demand, in 1979 the world was suddenly in the midst of a second oil shock.

Americans had grown accustomed to the high oil prices that persisted ever since the 1973 OPEC embargo. But the Iranian revolution made oil prices soar further, doubling over the course of a year to over $25 a barrel. This created havoc in America. Just as in 1973, truckers went on strike, parking their semis across roads and highways to block traffic. Faced with transport bottlenecks, farmers plowed their crops under. Long gas lines formed in California, and the state again used odd-even rationing, based on cars' license plate numbers. The gas lines, and then rationing systems, soon spread across the country. Compared with the first oil shock, this second one saw far longer gas lines—and during customers' long waits, tempers often flared and fistfights broke out. A country tune, "Cheaper Crude or No More Food," called for cutting off food exports to the Middle East until they brought oil prices down. Even before the song was released on vinyl, *The New York Times* reported, it became "a new national anthem for heartland America."

Many Americans thought the problems must have been the result of a conspiracy. In the spring of 1979, a Gallup survey found about half of Americans said the shortages were a hoax, a scheme to boost oil company profits. President Carter responded, "The energy crisis is real. I said so in 1977, and I say it again." That summer, as oil prices continued rising, another survey found that two-thirds of Americans called the crisis a hoax.

As the second oil shock continued to take an economic toll in the United States and worldwide, Carter's approval ratings sank below 30 percent—lower than Nixon's at the worst point of the Watergate scandal. Carter began to give increasingly morose, moralistic speeches, telling the public there were no "magic cures" for the energy crisis. "One of the most immobilizing fears among our people is the fear of being cheated and misled," he said. "This keeps people from making a small, personal sacrifice to solve the problems of inflation and energy."

To rally the nation, on July 15, the president made a televised speech on energy—one he'd obsessed over for weeks, retreating to Camp David and discussing the nation's problems with all manner of Americans, from ivory-tower academics to religious leaders to factory workers. When he finally delivered his speech, he opened by admitting, "I began to ask myself the same question that I now know has been troubling many of you. Why have we not been able to get together as a nation to resolve our serious energy problem?"

The underlying problem, Carter felt, was a widespread "crisis of confidence." Citizens were losing faith in a pillar of the American dream, that one's children would have better lives than they did—that is, of unending progress. Americans were becoming self-indulgent, building identities around owning and consuming things. But many were realizing that this way of life "does not satisfy our longing for meaning," he said, and meanwhile they were also coming to a new understanding of resources. "We believed that our nation's resources were limitless until 1973," he said, "when we had to face a growing dependence on foreign oil."

Through Carter's term in office, US oil production continued falling and imports remained near record highs, yet he held out hope that the nation could slash its dependence on foreign oil. This would require major efforts in energy conservation, as well as a crash program to develop oil shale—the huge resource in the Rocky Mountains that had yet to become commercially viable, despite high oil prices. Inspired by successes during the Second World War, Carter called for a National Energy Mobilization Board to steer policies—a technocratic approach.

"There are no short-term solutions to our long-term problems. There is simply no way to avoid sacrifice," Carter concluded. "Let us commit ourselves together to a rebirth of the American spirit. Working together with our common faith we cannot fail."

After the speech, a television host on ABC called it "almost a sermon." Carter's approval ratings soared.

Days later Carter announced a shake-up in his cabinet. He had fired half of them—including Schlesinger, the energy secretary, whom Carter had once called his "most important appointment." Schlesinger had led the administration's push for new energy policies, but that meant he had to serve as the promoter of unpopular policies. One of the most fraught was the proposal to scrap limits on oil and gas prices, which meant— at least for the short run—that Americans would have to pay more for energy. It didn't help that Schlesinger was plainspoken, arrogant, and often abrasive. He'd made many enemies on Capitol Hill and in industry. Carter felt he had little choice but to let Schlesinger go.

Rather than seeing the cabinet firings as a renewal, many took it as a sign of a crisis within the White House. The media and public soon soured on the president again.

"The depth of our national problem has not as yet been accepted by the American people," Schlesinger wrote in his letter of resignation from the administration. "This nation will, during the decade of the 1980s, face shocks with regard to energy supplies, for which we are not fully prepared as yet."

Compared with Hubbert's outlook on world oil, Schlesinger's was more pessimistic—but it was based in part on Schlesinger's reading of geopolitics. Saudi Arabia, for example, had been talking of the corrupting influences of oil revenue on its country and of cutting its production from around 9 million barrels a day to only 3 million.

Such differences aside, Schlesinger appreciated Hubbert's forecasts. "I was going to give Hubbert a medal," Schlesinger recalled, "to publicize what the long, long problem was for the country." But before he followed through on this plan, he'd been fired.

IN DECEMBER 1979, a year after the Iranian revolution, the Soviet Union invaded Afghanistan. Only a few hundred miles of desert in eastern Iran separated the Russians from the gateway for Persian Gulf oil, the Strait of Hormuz, dubbed "the greatest economic choke point in the world."

A month after the Soviet invasion, Carter declared in his State of the Union speech, "Let our position be absolutely clear: An attempt by any outside force to gain control of the Persian Gulf region will be regarded as an assault on the vital interests of the United States of America, and such an assault will be repelled by any means necessary, including military force." Although the Nixon and Ford administrations had flirted with war over Middle East oil, the United States had never formally stated such a policy. The new stance soon became known as "the Carter Doctrine."

In addition to the Soviet threat, oil markets faced other obstacles. A bloody war between Iran and Iraq held oil production down in both nations, helping push prices above $35 a barrel. It was a record high, and oil became a major issue in the 1980 presidential campaign. Carter vied against Ronald Reagan, who maintained his stance that America had vast resources and could extract plenty—if only the government would "get the hell out of the way."

In his standard stump speech Reagan claimed, "We learned from the US Geological Survey that there is more oil in Alaska than there is in Saudi Arabia." Journalists contacted the survey and found Reagan was far off the mark. "When it comes to US oil reserves," the *St. Petersburg Times* reported, "there is no escaping the conclusion that Ronald Reagan is either (1) ignorant, (2) a liar, (3) muleheadedly stubborn."

When it came time to vote—with the US economy still in the dumps, suffering from a recession, high inflation, and high unemployment—Reagan's sunny optimism appealed more than Carter's gritty realism. Reagan easily won the election.

Heroic Dreams

SOON AFTER REAGAN'S VICTORY IN the presidential race, the nation's skepticism about shortages, and its renewed optimism in markets and technology, were reflected in a *60 Minutes* television special on energy, titled "What Energy Crisis?"

The show frustrated Hubbert. It was "as misleading as it is possible to be," he wrote to a friend. When it came to natural gas, he said, "Great emphasis was placed on unconventional sources, especially the so-called geopressured gas in coastal Louisiana and Texas." (This was the resource McKelvey and *The Wall Street Journal* had hailed as holding enough for hundreds of years.) "Even if this gas exists, there is as yet little evidence that a significant amount of it will ever be produced," Hubbert argued. "There is also lots of gold in the oceans."

Most resources in the ground—both metals and fossils fuels—would likely remain there because they were too diffuse to be worth extracting, Hubbert figured. Decades earlier the German chemist Fritz Haber had run up against such difficulties. In the early 1900s, Haber had figured out how to pull nitrogen out of the air and create fertilizers and explosives, turning what had been a limited resource into an abundant one—a feat that won him a Nobel Prize. After World War I, when the victors levied Germany with heavy reparations, Haber hoped to conjure up enough gold to pay off the victors. He embarked on a mission to extract gold from ocean water. He did manage to muster some small ingots. However, despite his obsessive efforts, the process proved so difficult that it wasn't worthwhile.

More generally, Hubbert thought society couldn't simply pay more and more and somehow unlock infinite resources. It irked him whenever others talked of rising oil prices as a panacea that would enable companies to finally tap resources they hadn't been able to profitably extract before. When this argument came up at a conference on forecasting techniques, Hubbert sounded off. While agreeing that higher prices would allow companies to extract more oil, "the effect may easily be exaggerated," he said, adding:

> If oil had the price of pharmaceuticals and could be sold in unlimited quantity, we would get it all out except the smell.
>
> However there is a different and more fundamental cost that is independent of the monetary price. That is the energy cost of exploration and production. So long as oil is used as a source of energy, when the energy cost of recovering a barrel of oil becomes greater than the energy content of the oil, production will cease no matter what the monetary price may be.

"THAT MAN WAS so pessimistic that it just created a pessimistic idea that we weren't going to find any more oil and gas," Houston oilman Michel Halbouty told a reporter in 1981. "That man did more damage to the thinking of Congress and this country than any one man I know."

"That man" was Hubbert.

A petroleum geologist who had made a fortune, Halbouty liked to issue his own oil forecasts. But like those of the ancient oracles—whom Hubbert criticized for "prediction by ambiguous statement"—Halbouty's prognostications were usually vague. "We're going to find oil and gas way into the 21st century," Halbouty said on one occasion. Other times he was somewhat more specific, saying, "There is as much oil to be found in the United States, and the world, as has been found to date."

Halbouty's optimism meshed with President Reagan's. "Reagan and I used to talk a lot," Halbouty recalled. "I was his energy guru." The two men—the geologist-turned-guru and the actor-turned-politician—shared the view that solving the nation's energy predicament was simple. Regulations had put a cap on the price for "old oil," extracted from existing fields, so it sold for a fraction of world market prices. (Companies were allowed to charge whatever they wanted for "new oil" from recently discovered reservoirs.) Carter had begun to phase out those price controls, but Halbouty wanted to end them immediately—and this, he thought, would make it profitable to use much more intensive methods to get oil out of the ground. With the end of price controls, Halbouty argued, newly free markets would supply an abundance.

The industry distinguished different levels of effort for recovering oil. Primary recovery was simple: drill a hole in the right place, and oil would shoot up to the surface. Later came secondary recovery: techniques including waterflooding and gas pressurization, in which pumping water and natural gas into oil reservoirs kept the oil flowing.

By the early 1980s there was increasing talk of going further with efforts to pull recalcitrant oil out of the ground, which became known as tertiary recovery. Carbon dioxide—the main component of smokestack and tailpipe exhaust—could be pumped into reservoirs to dissolve the oil, helping wash it out of the reservoir. Or, to get viscous "heavy oil" flowing, steam and chemicals could be injected into underground deposits. Such methods had been used occasionally but were typically far more expensive than primary and secondary recovery. With widespread use of tertiary recovery, Halbouty thought, the United States could produce vastly more oil. "We've just got to go after tertiary recovery," he said. "It's a must."

Halbouty led Reagan's energy transition team, and in preparation for Reagan's inauguration, the team drafted an executive order to immediately remove all limits on oil and natural gas prices. Halbouty hoped Reagan would sign this order "one hour after he gets into the White House."

. . .

AMERICA WAS IN a bad spot, as even President Reagan admitted in his inaugural address in January 1981. "These United States are confronted with an economic affliction of great proportions," he opened. "We suffer from the longest and one of the worst sustained inflations in our national history," with a rate over 10 percent.

But Reagan held out great hope. "It is time for us to realize that we're too great a nation to limit ourselves to small dreams, " he said. "We're not, as some would have us believe, doomed to an inevitable decline. . . . We have every right to dream heroic dreams."

Reagan then pointed a finger: "In this present crisis, government is not the solution to our problem," he declared. "Government is the problem." The way to heal the country's wounds was through economic growth. "It is time to reawaken this industrial giant," by cutting taxes and the size of government. "These will be our first priorities, and on these principles there will be no compromise."

Reagan didn't approve the plan for oil price decontrol within an hour of entering the White House, as Halbouty had hoped. It took Reagan eight days. Still, it was one of his first steps, as he put it, to "get government off our backs."

"IT'S HARD TO believe, but some people continue to say that America has 'peaked,'" said one of the advertisements for "Growth Day," a new holiday promoted by big businesses. As a rejoinder to Earth Day, which had been running each April 22 for the past decade, Growth Day would preempt it, occurring on April 17. Major newspapers—*The New York Times*, *The Washington Post*, the *Los Angeles Times*, and the most pro-growth of all, *The Wall Street Journal*—ran Growth Day ads that declared: "Our national wealth isn't limited by our national resources. If growth has any limits at all, it's our personal resources—our drive, our dedication and our desire to lead better, more fulfilling lives."

This promotion of Growth Day, Hubbert wrote to a friend, "demonstrates much more vividly than I can the mental processes of the devotees of the monetary culture." He had just gone through a cataract surgery, limiting his ability to work. But what energy he did have, he devoted increasingly to his critique of "the monetary culture"—his name for a way of thinking, favored by economists, that measured everything in terms of dollars and ignored physical limits. When Hubbert received an invitation to speak at MIT, he crafted a talk similar to one he'd delivered in Berkeley several years earlier. This time he titled it "Two Intellectual Systems: Matter-Energy and the Monetary Culture."

The greatest challenge for the scientific world, Hubbert argued in his MIT talk, was to replace monetary accounting and thinking. In its place, he hoped the world would devise some other way of managing resources that would fit with physical realities—including limits to growth. Only such a system could run in a "near optimal manner," he argued. He was reviving Technocracy's old goal, aiming to supply the best possible life for everyone, given the available resources.

Since energy consumption underlay all activity—whether on farms or in factories—this new accounting system would have to represent the energy required for doing anything, from growing a head of lettuce to building a car. He argued for a new currency that would have "none of the properties of money" and "would be non-negotiable and hence not amenable to theft or other undesirable legal or illegal activities intrinsic in the properties of money." It was the same notion of energy certificates that Howard Scott had argued for in the 1930s.

Such changes would entail an overhaul of modern society, Hubbert thought, but this could enable people to have far more leisure time to devote to study, research, and reflection. "Were an inquiry of this kind to be successful," he concluded his talk, "it could well produce one of the greatest intellectual and scientific revolutions in human history."

Soon after that talk, Hubbert received surprising news from Columbia University. When he had taught there in the 1930s, his colleagues never backed his ideas for embracing new methods in earth science, incorporating physics and drawing heavily on mathematics. He'd been frustrated at their "stupidity," as he put it. "My departure was by mutual consent," he recalled. "They had had enough of me and I had had enough of them."

After World War II, Columbia had created a pioneering geophysics lab, the Lamont-Doherty Geological Observatory, which practiced the kind of integrated approach to earth science that Hubbert had long advocated. By the 1950s and 1960s, that approach had finally taken hold in many universities and research centers, in part because of Hubbert's efforts through the National Research Council, the Geological Society of America, and other bodies. In recognition of Hubbert's trailblazing efforts, Lamont-Doherty awarded him the Vetlesen Prize. It was the highest award in the earth sciences—the closest award this field had to a Nobel Prize.

At the Vetlesen Prize ceremony in December 1981, the presenter— Barry Raleigh, a fellow geophysicist—celebrated Hubbert's many achievements, including his prescient oil forecasts. "Hubbert's estimates were in direct conflict with official US Geological Survey estimates in the 1970s. His estimates are now, I believe, generally accepted as correct," Raleigh said. "Being outspokenly correct when the conventional wisdom would have it otherwise may not win popularity contests, but the vitality and intellectual integrity of men such as King Hubbert are rare and precious qualities."

The next day, as part of the awards ceremony, Hubbert gave an hour-long lecture at Columbia. He could have rehashed his geological work, the main reason he'd won the award. Instead he gave his "Two Intellectual Systems" talk, running through his various resource forecasts and laying out his hope for replacing "the monetary culture."

To enable such an overhaul of society, Hubbert thought the first step should be education. "Fundamentally I'm an educator and always have been," he later recalled. "I was in Shell. I was with the government." When it came to getting across ideas about resources, "It's a problem of trying to educate the public to the state we're in," he once told a journalist. "We must view it over a long time span. This phase we're in of exponential growth is about over. We're entering into something new."

· · ·

AT THE TIME of the 1973 OPEC embargo, the United States had around 1,500 drilling rigs tapping oil and gas. After prices soared in that first oil shock, the number of active drilling rigs had steadily increased, reaching around 2,200 by the time of the Iranian revolution in early 1979. Then, with the second oil shock, drilling had truly skyrocketed, reaching more than 4,500 active rigs at the close of 1981.

All this drilling, however, had failed to create a US oil renaissance. In the contiguous United States, production continued to decline at a rate of about 3 percent per year. With the addition of the giant Prudhoe Bay field, Alaska's oil production soared to 1.6 million barrels a day, enough to offset the decline from the rest of the United States, but the national total remained below its 1970 peak. Also, the Prudhoe Bay field was looking like a fluke. Even after a decade-long exploration boom in Alaska, the industry hadn't managed to find any more fields anywhere near as large.

Elsewhere, though, production was booming. Spurred by high oil prices, production ramped up quickly from huge offshore oil fields in the North Sea—divided between the United Kingdom, Norway, and Denmark—reaching about 2.5 million barrels per day by 1981. Production also soared at Mexico's Cantarell complex, a set of huge offshore oil fields.

Meanwhile, with oil prices persistently high and economies around the world in recession, the global appetite for oil had fallen about 15 percent from its peak just before the second oil shock. Efforts to cut oil consumption, with less use in power plants and more efficient cars, were paying off.

By late 1981 many experts declared an oil glut. Most OPEC nations stuck together, holding their prices steady and decreasing production to try to avoid flooding the market. But other major exporters—the United Kingdom, Nigeria, Mexico—competed in a shrinking market by lowering their selling prices, triggering prices worldwide to slide.

With this oil price drop, America's decade-long drilling boom quickly deflated. In the first nine months of 1982, the number of active drilling rigs plummeted by nearly half, from around 4,500 to 2,400. The synthetic fuel industry—which aimed to derive gasoline from coal and oil shale—also went bust, before it mustered any significant production.

While this drilling crash was bad news for drillers, most greeted it as good news, since lower oil prices would help consumers. Also, US oil imports were declining, and with prices falling, the amount Americans spent on imported oil was falling even faster. The price drop also seemed to signal that OPEC's control over world oil markets had been broken. Many experts announced the end of the long energy crisis.

However, Daniel Yergin of Harvard Business School countered, "We'd be making a terrible mistake to say our energy problems are over." Just thirty-five, Yergin was a rising star in academia. Having trained as a historian with a focus on the Cold War, in the late 1970s he had turned to focus on energy. He'd cowritten a wonkish book, *Energy Future: The Report of the Energy Project at the Harvard Business School*, which appeared during the second oil shock and became a surprise best seller. By the early 1980s, Yergin had become one of the nation's most widely quoted energy analysts.

"We don't have to go from the extreme of panic to the extreme of complacency," Yergin argued in the summer of 1982, as oil prices continued to fall. The nation had yet to address "the critical question of whether the United States can halt the decline in its oil production that began in 1970," he wrote in *The New York Times*. He lamented how motivation was evaporating for spending on conservation and alternative energy sources, and how projects across the board—from clean solar to dirty oil shale—were being canceled. This slackening of effort could come back to bite America and the world. "There is high probability of another oil shock, which could trigger far graver economic and political consequences," he wrote. Such a situation could lead to recession, unemployment, and racial strife so divisive that "the basic legitimacy of our political system might be called into question."

As the oil glut grew larger, some talked of "an avalanche of supply," but Yergin pointed out that the picture was not so rosy. Texas was the most gung-ho oil state, with more than a third of the nation's active drilling rigs—yet the state's oil production continued to decline from its 1972 peak. "The tragedy for Texas," Yergin said, "is that it's locked into an industry with a declining production curve." Michel Halbouty and many others had touted enhanced recovery as the next big thing, but it had failed to turn around America's oil decline. So, Yergin argued, "it's not a good idea to simply repeat the words 'enhanced recovery' as a soothing mantra."

In their outlooks for the future, Yergin argued, oil analysts had gone overboard. "They assume that everything that is announced will be built, that all expectations will be met, that problems and confusion won't get in the way and that everybody will have plenty of capital to invest in conservation and alternatives—in short, that everything will turn out for the best."

AS OIL PRICES continued to fall through 1982, Yergin argued many had become seduced by "wishful thinking" that oil prices could drop as low as $15 a barrel.

Hubbert, though, had continued to focus on the long term, whether prices were high or low. He agreed that higher prices would encourage people to produce more oil—or at least, to extract it sooner than they otherwise would have. But he still saw money mainly as a bookkeeping device and thought changing prices would have little impact on the long-term picture. As he'd written to a colleague when prices were near their peak, "No monetary manipulation can find oil that isn't there."

Hubbert felt this point was captured best in a cartoon by Ed Stein that appeared in the *Rocky Mountain News*. It showed a rifle-wielding pioneer talking to a Native American. "Buffalo shortage? *What* buffalo shortage?" the pioneer says. "Just give me more money for guns and scouts, and I'll get you all the buffalo you need." Buffalo had once dominated the prairie but were then driven nearly extinct by hunters, their numbers never to rebound.*

---

* Ironically, the US Department of the Interior, charged with overseeing the nation's resources, has on its official seal a single animal: a buffalo.

Through the oil-market roller coaster of the 1970s and early 1980s, Hubbert had stuck with much the same outlook he'd held for decades. As two Exxon analysts put it in 1982, Hubbert was "the only energy analyst we know of who today can use, to effect, charts he prepared over 35 years ago." And Hubbert did continue presenting his old graphs, which showed that production from the contiguous United States tracked his predictions fairly closely—despite the advent of OPEC control over markets and the rise of prices, despite the end of US price controls on oil and the global oil glut.

Hubbert also had a graph of world oil production, which he'd created in the mid-1970s and continued presenting in the following years—the same one the EIA had drawn on in its 1978 annual report. The graph showed two scenarios for world oil: one, a high peak in the 1990s with a sharp decline afterward, and the other a long plateau from the 1970s through the 2030s, followed by a sharp decline. Actual production would probably wind up somewhere between these two extremes. But where?

Those two extreme scenarios were both based on an assumption about how much oil the world might ultimately extract—much the same method Hubbert had first presented in 1956. Following that 1956 forecast, oil estimates had soared, showing how flexible they could be. So he'd become skeptical of such resource estimates and tried to avoid them whenever possible. Not content to keep showing his old graphs, he wanted to try to develop a better approach to forecasting the future of world oil.

Hubbert had been thinking for years about how to improve forecasts of world oil. But he started working on it only after being approached by James MacKenzie, a physicist working for the Union of Concerned Scientists, who suggested they collaborate. In his US studies, Hubbert had been able to get data that showed how many wells had been drilled and how much oil that drilling had discovered. When it came to world oil, data was "much scarcer," Hubbert told MacKenzie. "But I am sure that enough can be assembled to do a pretty good job." Soon after starting this collaboration, Hubbert wrote to a colleague, "I am beginning to assemble the most extensive and up-to-date data upon world production, proved reserves, and exploratory drilling that I can find, with the intention of doing an analysis similar to what I have done previously for the United States."

After several months, however, Hubbert and MacKenzie hit a wall. Drawing on the usual public data sources such as the American Petroleum Institute and the Department of Energy, there wasn't enough to go on. As Hubbert told another analyst, they were "stymied for lack of essential statistical data."

Then Hubbert found that an industry analyst he knew had a connection with Petroconsultants, a firm based in Geneva, Switzerland, that maintained perhaps the most comprehensive database on worldwide oil. Petroconsultants got proprietary data directly from oil companies, which shared the information because they felt it was in their interests to have it compiled by a trusted third party, allowing the companies to track how the industry as a whole was developing.

Petroconsultants agreed to let Hubbert use some of their data, and by the fall of 1983, he and MacKenzie began receiving tables of numbers. But what they got fell short of what Hubbert wanted. Lacking information on reserve growth over time, he couldn't do the kind of comprehensive analysis he'd done in his 1967 paper, "Degree of Advancement of Petroleum Exploration in United States." He was stuck again. For years, he'd been unsatisfied with the rough methods he'd been using for assessing world oil prospects—but it wasn't easy to do a better analysis.

Before managing to get a hold of more data, Hubbert would suddenly find himself unable to work. A more in-depth study would have to wait.

Things We Could Do *Tomorrow*

IN JANUARY 1984, AT AGE 80, King Hubbert underwent a minor operation to repair a hernia. Afterward he suffered breathing problems. Initially diagnosed as pneumonia, it was a pulmonary embolism—a blockage in an artery in his lung. If the clot got loose and wound up in his heart or brain, it could kill him. The embolism put him in the hospital for an extended stay for the first time in his life.

Even when back home, his doctors had him on an anticlotting medicine for several months. The drug left him "mentally semi-deranged," as he put it. One day when Miriam was shopping at the supermarket, she noticed that a common rat poison contained the same chemical King was taking.

Throughout 1984—while Hubbert rested and was unable to work— oil prices continued to tumble, falling below $30 a barrel for the first time since the Iranian revolution. Cheaper oil had "strengthened what looked to be a fragile recovery," Daniel Yergin told *The New York Times*, "as much a reason for the current economic recovery as President Reagan's tax cuts." While Nixon and Carter had fought headwinds of rising oil prices, Reagan enjoyed a tailwind of falling oil prices.

With the economy on the mend, Reagan sailed to reelection in November 1984. His second inaugural speech soared higher than the first.

"There are no limits to growth and human progress," Reagan declared, "when men and women are free to follow their dreams."

Hubbert, however, bemoaned "the deplorable and ominous state of the country under the Reagan administration." Reagan slashed spending on alternative energy, which Hubbert called "sabotage" of solar energy. Reagan did cut taxes for some and slashed many parts of the government. But Reagan also hugely boosted military spending, increasing the nation's debts. Hubbert saw such military buildups—along with deficit spending by the government, and consumers piling up debt—as part of a pattern for boosting economic growth, one he'd been lamenting for decades.

When a friend sent King and Miriam the book *World Military and Social Expenditures*, detailing many nations' high military spending relative to other investments, such as for education and health, Hubbert replied, "It confirms in devastating detail what we already knew in a general way is occurring in the world. The situation is one of complete madness, and I can see little hope for bringing it under control."

BY MID-1985, HUBBERT had been "essentially *hors de combat*" for a year and a half. Not one to gripe about his aches and pains, nonetheless Hubbert was clearly frustrated at being unable to work. But he did keep in touch with a few friends, in particular other like-minded researchers. That June one of his friends sent him Chevron's latest annual *World Energy Outlook*. At the time, the Chevron report explained, there was "a general complacency about the adequacy of future oil supply." The report wanted to counter that view, arguing that there actually were limits in sight for conventional oil.

In gauging the world's ultimate production of crude oil, the report said, "2 trillion barrels is a value which has been widely accepted for a number of years, and which is bracketed by credible recent estimates." That amount—2,000 billion barrels—had been, for more than a decade, Hubbert's favored estimate for the ultimate potential for world oil. By the 1980s, Chevron argued, the petroleum industry had come to the same view.

"At some time early in the next century," the Chevron report continued, "conventional oil production can reasonably be expected to reach a peak and then begin a long-term decline." That agreed with other industry forecasts, such as Exxon's 1980 *World Energy Outlook*, which had noted that oil discoveries were on the decline, to the point that companies were finding less oil than they were extracting. "Since production cannot increase indefinitely in the face of declining discovered reserves," the Exxon report had said, "it seems reasonable to expect that conventional oil production will reach a plateau some time shortly after the turn of the century." Chevron's 1985 report extended such forecasts out much further, showing conventional oil production peaking around 2005—and after that a long decline, dwindling to almost nothing by 2100.

Delaying the peak significantly would require enormous new discoveries, the Chevron report explained, using language that echoed Hubbert's 1956 study "Nuclear Energy and the Fossil Fuels." "An additional 100 billion barrels," Chevron said, "could defer the start of the decline by less than 5 years or support only a modest increase in the peak rate."

For decades, the oil industry had maintained that with "adequate economic incentives"—high enough prices and low enough taxes—it could supply as much fuel as the nation wanted. Hubbert called this the industry's "persistent propaganda line." But the nation had seen high prices ever since the early 1970s, and Reagan had lifted price controls in the 1980s—yet US oil production continued to fall well short of US consumption.

With limited prospects for conventional oil in the US, many companies had turned their sights to enhanced recovery and unconventional oil. These unconventional sources had long been known. Back in the 1950s, Hubbert had expected many alternatives could replace oil. Coal could be turned into liquids, as could tar sands and oil shale. But in the three decades since, he'd soured on those options. The costs—in terms of dollars, in terms of environmental impact, and in terms of energy inputs required—were too high, he believed, for these alternatives to make more than a modest contribution.

This is where the oil industry disagreed with Hubbert's outlook. In the early 1980s, Exxon had expected that unconventional oil production would soar, supported by higher oil prices, and was at the time sinking large investments into an oil shale project in Colorado.

Chevron, too, was hopeful about unconventional oil. It was one of the biggest players in enhanced oil recovery, especially from California's fields of viscous "heavy oil." So Chevron argued that, over the long run, oil prices would rise, which would discourage people from consuming as much and at the same time open up new possibilities for enhanced recovery and synthetic oil. With higher prices, "the available supplemental resources are immense," Chevron argued, able to provide a "large but expensive supply" of fuel "for many years."

After oil prices started declining in the early 1980s, Exxon and other oil companies had pulled out of their oil shale projects, despite having sunk hundreds of millions of dollars into these efforts. Exxon's president explained that costs for oil shale extraction had continued rising, so "the final cost would be more than twice as much as we thought it would be when we entered the project." The president of another company involved, Tosco Oil, said oil shale's financial failure showed "the need for Government participation on a sustained and intelligent basis in order to make such a field demonstration on a commercial scale fully effective."

Likewise, Chevron's 1985 report took a pessimistic turn. But its message did align with the company's interests at the time.

Chevron, like other companies, hoped some combination of higher oil prices and subsidies would make many unconventional oil projects viable. The catch was that people would have to pay substantially more for fuel. The world had already been through two oil shocks, both of which triggered deep recessions in the United States and encouraged people to use oil more efficiently. But in Chevron's presentation, the high price tag on alternative fuels was no obstacle.

CONTRARY TO CHEVRON'S hopes, oil prices continued to drop through the summer of 1985. Daniel Yergin—by then cofounder and president of a consulting firm, Cambridge Energy Research Associates— issued another warning that these low prices were feeding complacency.

"Unless there is a major technological development," he wrote in *The New York Times*, "at some point the reduction in energy investment will come back to haunt us, and market realities will again give way to geological realities—the concentration of oil reserves in OPEC and in the Middle East. And that will eventually put the era of surplus behind us."

However, soon Saudi Arabia—the pillar of OPEC and the world's largest oil exporter outside the Communist bloc—announced a major policy shift. For years, Saudi Arabia had served as the world's "swing producer," adjusting its production up and down as needed to keep oil prices relatively stable. When prices had been at their peak in the early 1980s, Saudi Arabia produced over 10 million barrels a day. But as oil prices fell, it continued slashing its production until, by July 1985, it was down to 2.2 million a day—and yet oil prices continued to fall. Saudi Arabia could cut no more.

Through 1985, as OPEC members struggled to come to agreement on how much they'd charge for oil and how much they'd produce, Saudi Arabia threatened to quickly double its production to 4 million barrels a day—and possibly raise it to 9 million by the end of the year. To earn enough to support its petrostate, Saudi Arabia would flip from low volume and high price to selling more barrels cheaply. "It may be risky for them to boost production and force prices down," *The New York Times* noted, since it meant OPEC was essentially giving up its control over world oil prices. Things could be different a decade down the road, however. "By the 1990's, when the rest of the world's oil production is expected to decline," the *Times* added, Saudi Arabia "could be back in the driver's seat."

In early 1986, Saudi Arabia did raise its production and oil prices went into free fall. A few years earlier Yergin had lambasted others for "wishful thinking" that prices could drop to $15 a barrel—but by February 1986, that had happened. By the summer, some selling prices—such as the Brent benchmark for North Sea oil—were approaching $10 a barrel. America's oil exploration boom, which had taken a nosedive a few years earlier, truly collapsed. In Dallas, a bumper sticker became hugely popular: "Please Lord, Give Me One More Boom—This Time I Promise Not to Piss It Away."

Interest in energy issues also went bust. For more than a decade, Americans had been urged to cut back—to turn down thermostats, buy smaller cars, and invest in home insulation—all to avoid worse crises in decades to come. "All that is filed away now," the *Los Angeles Times* lamented, "as part of yesterday's problem." Gerald Greenwald, the chairman of Chrysler Motors, one of Detroit's Big Three car manufacturers, said, "Gasoline is flowing as freely today in America as beer at a fraternity party."

While most automakers welcomed low oil prices, the new attitude was dangerous, reflecting "much the same short-sightedness that set America up for the oil shocks of 1973 and 1979," Greenwald argued in a *New York Times* opinion piece. The nation had established CAFE standards, he pointed out, requiring carmakers to produce vehicles that averaged at least 27.5 miles per gallon. But soon after the drop in oil prices, there were calls for scrapping these efficiency requirements. The nation risked repeating the mistakes of the past, Greenwald argued: "Each time, Americans wanted to know how Detroit—and the Government—could have been so stupid. Why weren't we prepared? Why did our cars consume twice as much gasoline as necessary?"

The nation suffered from an ongoing problem of a short-term view and a lack of planning, Greenwald wrote. When the earlier oil shocks hit, "we had no national energy policy to temper wild swings in energy supplies and prices. We still do not." He concluded, "Washington should have learned a lesson from the first two energy crises. I don't know what ought to be said about a country that lets itself get hurt three times in a row."

The oil industry typically pursued less regulation, not more—but with the oil price crash, some considered desperate measures. In late 1986 George Keller—CEO of Chevron as well as chairman of the American Petroleum Institute—called on the government to aid US oil extraction by establishing a minimum price for oil (the role the Texas Railroad Commission had played before US oil production peaked). "From the standpoint of America's economic and energy security," Keller said, "our nation cannot afford an extended disruption of domestic drilling."

Keller called for the industry to form a united front in pushing the government to help—and to overcome Reagan's aversion to regulations. Others in the industry were skeptical, but such regulation did have the support of one high official: the vice president, George H. W. Bush. Observing the industry's about-face on regulation, a former energy official commented, "Ideology always takes second place to money."

BY 1986, HUBBERT'S acute problems from his embolism had faded. But he still faced setbacks that prevented him from doing much work. He caught a flu and, a month later, was still "so disorganized that I can barely add 2 + 2," he told his sister Nell. By the summer, as the oil price collapse bottomed out, he finally got well enough to start work again.

"I am still following the evolution of the US and world petroleum industries with great interest," he wrote to a friend that summer. "I find myself humming a parody of the lovely song, 'Where have all the flowers gone?'" His version went:

Where have all the oil fields gone?
To depletion, every one.
When will they ever learn?
When will they ever learn?

Throughout his life he'd been concerned about humanity running short of fossil fuels—but in the 1980s he also grew concerned about burning too much fossil fuels. He read a ground-breaking paper by researchers at NASA, "Climate Impact of Increasing Atmospheric Carbon Dioxide," published in *Science* in 1981. The study warned that burning fossil fuels had released enough carbon dioxide to cause significant global warming—warming that would continue increasing and would soon become large enough to stand out from natural fluctuations in Earth's temperature. That is, after decades of hypotheses about man-made climate change, it would soon become undeniably real.

Continued fossil fuel consumption, the study estimated, could easily raise global temperatures around a few degrees Celsius by the year 2100—enough to melt glaciers around the world as well as the West Antarctic ice sheet, eventually leading to a sea-level rise of around 5 meters, or 16 feet. Hubbert underlined a passage in the study that this scale of sea-level rise "would flood 25 percent of Louisiana and Florida, 10 percent of New Jersey, and many other lowlands around the world."

Hubbert's outlook on the size of the resource base was roughly similar to what the NASA researchers foresaw. That is, even if the world's oil and gas production both peaked and declined during the twenty-first century, there would still be enough coal to disrupt the planet's climate. Such warnings about global warming gave an additional impetus to Hubbert's lifelong search for alternative, long-lasting energy sources to replace fossil fuels.

Nonetheless, into the late 1980s, Hubbert continued to oppose nuclear power. As he put it in a 1988 interview:

> We went into nuclear quite optimistically, including myself, but it turned out to be a very, very hazardous business. We're in the midst of a hell of a lot of trouble with nuclear power plants. We've got the waste disposal problem. We've got the associated atom bomb problem. We've got the problem of what are we going to do with these plants when they wear out. We haven't faced up to that problem yet. So my conclusion right now is if we had breeder reactors, that would prolong the life of our fuel supply, but all the other problems are still with us.
>
> So then we go to fusion. Every year about the budget time, we get these great glowing stories about how close we are to producing fusion. We never quite get there. We haven't got fusion, except for atom bombs. So let's just ditch nuclear energy for the time being and see what else we've got.
>
> Well the biggest source of energy on this earth, now or ever, is solar.

Hubbert imagined that solar plants could cover the deserts, generating electricity. The electricity could also be used to drive chemical reactions to form methanol, a liquid that could be sent through pipelines. Or the energy could be used to generate hydrogen, which would be another way of storing energy so it could be transported. When the hydrogen was used to generate energy, it would react with oxygen in the air, and "the combustion product is drinking water—no carbon dioxide, no fouling the atmosphere."

He continued:

This technology exists right now. So if we just convert the technology and research and facilities of the oil and gas industries, the chemical industry and the electrical power industry—we're not dealing with some hypothetical future, we're dealing with things we could do tomorrow. All we gotta do is throw our weight into it.

Were we a rational society—a virtue of which we have rarely been accused—we would husband our dwindling supplies of oil and gas, supplemented by imports as long as they are available, and institute a program comparable to that in the nuclear industry of the 1940s, 50s and 60s, for the conversion over to solar energy. These things we could do *tomorrow*. We could improve the technology, but we have the technology already.

His optimism about the technologies aside, Hubbert was still deeply worried about the "exponential growth culture" that was entrenched in the United States and increasingly so in the rest of the world. As Hubbert put it in that 1988 interview, "One of the most ubiquitous expressions in the language right now is growth—how to maintain our growth. If we could maintain it, it would destroy us."

ON THE EVENING of October 10, 1989, a week after King Hubbert's eighty-sixth birthday, he and Miriam went to the Cosmos Club, a favorite haunt, and met a friend for dinner. Hubbert looked well and was in great spirits. On returning home that night, he did some paperwork before going to bed. He never awoke. In the early hours of the morning, the embolism that had threatened him for years finally caught up with him.

Hubbert had requested cremation and to have his ashes sent back to San Saba, Texas. He'd left his boyhood home as soon as he could, at age seventeen, and returned only rarely for visits, but it was where he wanted to be buried.

On his death, *The New York Times* and *Washington Post* both ran obituaries that cited his wide-ranging geological work and hailed him as a prophet who'd warned the country about coming energy crises. His involvement with Technocracy and his broader ideas about transforming society, however, were either forgotten or ignored. But the *Times* obituary did feature an old photo from the mid-1930s that showed him scrawling equations on a chalkboard. On his lapel—a tiny detail impossible to discern in the newspaper's printed version of the photo—Hubbert was wearing a small pin with a yin-yang design: Technocracy's emblem.

EPILOGUE: 1990–2015

"M. KING HUBBERT should be spinning in his grave," the business magazine *Barron's* proclaimed.

It was 2012, and the United States was in the midst of a drilling boom unlike any it had ever seen. US oil production had been in a nosedive for decades, but it suddenly rocketed upward faster than ever before, making the nation far less dependent on imports. Other oil-importing nations—from the United Kingdom to China—viewed the US boom with envy. It was completely unexpected, and very welcome.

Buoyed by this wave of optimism, the investment bank Citi released a report titled "Resurging North American Oil Production and the Death of the Peak Oil Hypothesis." High oil prices had enabled this turnaround, making it attractive for companies to use intensive extraction processes they hadn't employed before. However, Citi argued, the high prices were not due to any actual shortages. "The belief that global oil production has peaked, or is on the cusp of doing so," its report said, "has helped to fuel oil's more than decade-long rally."

Citi's belief, on the other hand, was that the world would soon enter a new era of significantly cheaper oil.

CHEAP OIL WAS something the US was willing to fight for. From Schlesinger and Kissinger during the 1973 OPEC embargo, to the Carter Doctrine after the second oil shock, and into the 1980s, the United States' increasing dependence on oil imports made the option of military intervention a recurring theme.

In 1990, when Iraq invaded Kuwait in a dispute over oil fields along their border, the United States quickly took action. US forces moved into Saudi Arabia to defend that oil-rich neighbor of Kuwait, and to drive back Iraqi forces. In telling the American people of his decision to launch this war, President George H. W. Bush began by invoking an "age of freedom," of the need to resist aggressors and stand up for principles. But he also spoke of America's "vital interest" in the Persian Gulf. "The stakes are high," Bush said. "Our country now imports nearly half the oil it consumes and could face a major threat to its economic independence. Much of the world is even more dependent upon imported oil and is even more vulnerable to Iraqi threats." Since the war might disrupt the flow of oil from the Persian Gulf, Bush added, "I will ask oil-producing nations to do what they can to increase production in order to minimize any impact that oil flow reductions will have on the world economy."

New York Times columnist Thomas Friedman put it more plainly. "Laid bare," he wrote, "American policy in the gulf comes down to this: troops have been sent to retain control of oil in the hands of a pro-American Saudi Arabia, so prices will remain low."

As the United States prepared for war, oil prices spiked—and the US economy sank into a minor recession. The war—in which the United States pushed Iraq out of Kuwait—wound up knocking down Iraqi production from 3 million barrels a day to under half a million. Iraq remained under US sanctions through the 1990s and its oil production recovered somewhat, but not to its prewar levels. Total world production nonetheless continued rising about 1 percent per year. Oil remained cheap, and international trade boomed. The buzzword of the decade was *globalization*. There was no longer much discussion of the old worries about ever-higher oil prices and an approaching peak of world oil production.

Nonetheless, a few experts warned there was still reason for concern. Two retired industry veterans—Colin Campbell, who'd spent his career at British Petroleum, and Jean Laherrère of Paris-based Total—warned in a widely read 1998 *Scientific American* article, "The End of Cheap Oil," that the world's conventional oil production would likely peak within a decade. A similar prediction came from Kenneth Deffeyes, Hubbert's former colleague from Shell's Bellaire lab, in his book, *Hubbert's Peak: The Impending World Oil Shortage.*

Oil prices began rising in 2000, becoming a major issue in the US presidential elections, since many Americans considered prices at the time—around $30 a barrel—to be onerous. Texas governor George W. Bush, running on the Republican platform, promised he would "convince our friends at OPEC to open the spigots," adding, "that's what diplomacy is all about." Bush's opponent, Vice President Al Gore, also hoped to bring oil prices down, but by different means. "I am going to keep challenging big oil," Gore said, "until America gets some straight answers and affordable gas prices they deserve."

Bush won the election and in 2002 began a push to invade Iraq because Saddam Hussein allegedly had weapons of mass destruction. Administration officials roundly rejected the notion that the war was motivated by oil. As Secretary of Defense Donald Rumsfeld said on a radio show, "It has nothing to do with oil, literally nothing to do with oil."

A few did break ranks, however. As the US was preparing the invasion, White House economist Larry Lindsey told *The Wall Street Journal*, "When there is a regime change in Iraq, you could add three million to five million barrels of production [per day] to world supply." He added, "The successful prosecution of the war would be good for the economy." After the invasion went forward in 2003, Alan Greenspan, the longtime chairman of the Federal Reserve, reflected on the situation in his memoir, *The Age of Turbulence.* "I am saddened," he wrote, "that it is politically inconvenient to acknowledge what everyone knows: the Iraq war is largely about oil."

After the US overthrew Saddam, Iraq descended into civil war, with many attacks on pipelines and other parts of the oil system—the kind of "skulduggery" warned of in the 1975 report *Oil Fields as Mil-*

itary Objectives. Elsewhere, oil production was struggling to keep up with fast-rising consumption, in particular in booming China. Yet the world's most prominent forecasters, including the EIA, expected oil prices to remain roughly flat for decades to come.

Instead prices continued rising persistently. By 2004, as oil passed $40 a barrel, the oil industry began discussing the issue of peak oil far more openly. That year the Offshore Technology Conference, one of the industry's largest trade shows, held a session titled "Hubbert's Peak," and the industry's flagship publication, *Oil and Gas Journal*, ran a series on the topic titled "Hubbert Revisited." In a *New York Times* article, energy historian and analyst Daniel Yergin asked, "Are the peakists right?"

Back in the 1970s and early 1980s, Yergin had shared such fears about another oil shortage—one that might be worse than the two oil shocks of the 1970s. But then in the mid-1980s, a glut led to the oil price crash, and Yergin's attitude flipped. He became one of the greatest optimists about oil supplies. Over the years, he remained one of the nation's most influential energy analysts, having written a Pulitzer Prize–winning history of oil, *The Prize*, and being regularly quoted in the media and invited to testify in Congress.

When peak oil fears flared up in the mid-2000s, Yergin argued there was nothing to worry about. There had been several such scares over the decades, he pointed out, and every time the peak had failed to materialize. The latest spate of peak oil worries was "not borne out by the fundamentals of supply," he argued. His consulting firm forecast "a large, unprecedented buildup of oil supply," which would enable production to soar 20 percent in the following five years, passing 100 million barrels a day by 2010. "This forecast is not speculative," he stressed. "It is based on what is already unfolding."

A similarly optimistic view came from the International Energy Agency (IEA). Though established to serve as a watchdog to warn of coming oil shortages, through the early 2000s, as oil prices continued climbing, the IEA's annual *World Energy Outlook* reports were reassuring. In the introduction to a major 2005 report, IEA director Claude Mandil wrote:

Soaring oil prices have again spotlighted the old question. Are we running out of oil? The doomsayers are again conveying grim messages through the front pages of major newspapers. "Peak oil" is now part of the general public's vocabulary, along with the notion that oil production may have peaked already, heralding a period of inevitable decline.

The IEA has long maintained that none of this is a cause for concern. Hydrocarbon resources around the world are abundant and will easily fuel the world through its transition to a sustainable energy future.

Yet even as respected forecasters were saying there was nothing to worry about, the world's conventional oil production hit a plateau. The resulting price rise—relentless, year after year—was unlike any that had occurred in the history of oil.

This wasn't like the first oil shock, in 1973, when OPEC had cut back production to get its way. It wasn't like the second oil shock, in 1979, when Iran's production dropped off drastically during its revolution. The oil price spike of the 2000s had no obvious trigger, so there was no one to point the finger at. Around the world, oil production was simply failing to meet expectations.

WHEN FIRST RUNNING for the presidency, Bush had promised to convince OPEC to open the spigots, but it didn't—or at least not enough to prevent prices from soaring. By January 2008, oil hit $100 a barrel—a major psychological milestone and nearly a record high (in inflation-adjusted dollars). At that point, Bush arranged to visit Saudi Arabia to talk to the king. Beforehand, Bush told a reporter, "I will say to him, 'If it's possible, your majesty, consider what high prices are doing to one of your largest customers.'" It wasn't good for the Saudis, either, Bush argued: "The worst thing that can happen to an oil-producing nation is that the price of oil causes the economy to slow down,'" because then consumers would buy less oil.

After this private plea, the Saudi oil minister publicly rebuffed Bush, saying his nation would raise production "when the market justifies it." A few days later the US energy secretary Samuel Bodman said, "The economy has been able to withstand it until now. I believe the $100 price of oil is starting to have an impact."

But prices continued rising, spiking to $150 a barrel by the summer of 2008. At that point, Bodman remarked, "We're in a difficult position, where we have a lid on production and we have increasing demand in the world."

Oil prices finally dropped only when the US economy went into meltdown in the summer of 2008. This "Great Recession" then spread through Europe and Japan. In the next couple of years, as the world economy made a slow recovery, oil prices quickly bounced back up, again reaching $100 a barrel in 2011.

By that time, though, many Americans saw an upside in the high price of oil, because it fueled the fast-growing US oil boom. High prices created jobs, reversed the nation's long decline in production, and slashed oil imports—a development the IEA called "nothing short of spectacular." It was the boom that, some said, had proved Hubbert wrong—and should make him spin in his grave.

THIS NEW OIL boom had its roots in earlier booms in tapping natural gas, spurred by the shortages of the 1970s. After America's conventional gas production went into decline, government and industry alike began scrambling to tap various gas resources that could be opened up only with fracking. Used on limestone and sandstone, fracking unlocked what was called "tight gas." Coal seams could be fracked too, releasing coalbed methane. What these resources all had in common was that the rock was almost impermeable to gas—so when the rock was drilled, little gas escaped. But if they were also fracked, that opened up pathways for extracting profitable amounts of gas.

These sources got a big boost in the 1980s, when the federal government deregulated natural gas prices, allowing them to rise, and also created tax breaks for unconventional gas. Tight gas boomed, making a major contribution to the nation's total production, and coalbed methane played a supporting role. Together these two unconventional sources managed to counter the decline of conventional gas and even push the nation's total production up somewhat through the 1990s, although it still remained substantially below its 1970s peak.

But by the early 2000s, as conventional gas production continued to decline, these unconventional sources weren't able to fill the gap. The nation's total gas production began falling again. Even Daniel Yergin, having become an optimist about oil, saw little hope for a US natural gas revival. There had been no major discoveries of natural gas in recent years, "though not for lack of effort, if you look at industry spending," Yergin said in 2004, when testifying before US Congress's Joint Economic Committee. "There is strong evidence that simply adding more drilling rigs will not solve the problem, as it has in previous decades," he argued. "United States gas productive capacity appears now to be in permanent decline."

Meanwhile a few drillers had been testing out fracking on another kind of rock, known as shale. The most dogged was George Mitchell, founder and CEO of Texas-based Mitchell Energy, who focused on the Barnett shale, a formation underlying the area west of the Dallas–Fort Worth metropolis. In fracking's early days after the Second World War, companies had pumped napalm—jellied gasoline—into wells to build up enough pressure to fracture them. Since then drillers had used all kinds of fluids to see what would work in a particular locale—experimentation that Mitchell's company continued. "We tried every kind of fracturing material," Mitchell recalled. "Oil, propane, ethane, all sorts of things." Once the drillers made a mistake in mixing a gelling compound into the drilling fluid, so it wound up very watery. It turned out this fluid actually worked better for fracturing the rock—a technique they soon adopted and dubbed the "slickwater" frack.

Meanwhile, with improvements in oil field technologies, drill heads were able to bore down some two miles, turn the drill bit through an arc until it was going sideways, then continue tunneling horizontally for another mile or two. The drilling was precise. A drill operator could snake a horizontal well through a shale formation just fifty feet thick, adjusting the well's trajectory up and down to follow the contours of the rock, keeping the well within the most promising part of the shale through its whole path. (The drill heads used gamma-ray detectors to pick up faint radiation naturally released by the rock, allowing a drill operator at the surface to tell whether the bit was still in the middle of the shale or was straying outside of it.) With a long horizontal well in place, it could be fracked all along its length, enabling each well to yield much more gas.

As natural gas prices more than tripled from 2002 to 2005, hovering around $7 per million cubic feet and occasionally spiking over $12, finally shale gas looked attractive. Shale drilling boomed across the country in various regions or "plays"—the Fayetteville in Arkansas, the Haynesville straddling Louisiana and Texas, and the biggest of all, the Marcellus that lay under most of Pennsylvania and stretched into neighboring states. By 2007 the nation's total natural gas production showed signs of a revival, which Yergin celebrated, dubbing it the "shale gale."

The downside of shale gas was that, after fracking released a burst of gas from a well, the well's production then plummeted, typically falling by about half the first year and continuing to decline in the following years. But the best shale gas areas had wells that were so prolific, and so much money poured into the sector, that the shale boom exceeded all expectations.

The boom was so rapid in part because drillers attracted large amounts of investment and borrowed heavily, allowing them to drill much faster than they could have afforded using their own revenues alone. One of the most aggressive was Chesapeake Energy, which grew quickly as it began snapping up shale drilling leases, becoming the nation's second-largest natural gas producer, behind only Exxon. It grew so quickly by attracting investors' money—and building up more than $12 billion in debt.

Other companies took a similar tack, and the debt-fueled shale gas boom pushed production up rapidly, by 2011 surpassing the early 1970s peak, allowing the United States to regain a title it had held through much the twentieth century: the world's top natural gas producer. Earlier fears of gas shortages were erased. Whereas many had been planning that the United States would become dependent on imports from overseas, delivered by ship as liquefied natural gas (LNG), instead there was talk of the United States becoming an LNG exporter.

The US Energy Information Administration was slow to recognize the extent of the boom. As late as 2010, it forecast that shale would play a marginal role and that the nation's gas production would only hold flat for the coming decades. But then in 2011, the EIA forecast that the shale gas boom would push production up and up, at least through the 2030s. With each new edition of its *Annual Energy Outlook*, the forecast was higher. Most industry forecasts were around EIA's, or higher, and similarly ratcheted upward with each subsequent report.

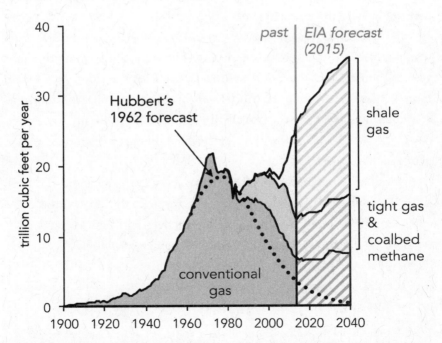

US natural gas production forecasts: For decades, the nation's conventional natural gas production closely followed Hubbert's forecast. However he underestimated the potential for unconventional gas. In 2015 the EIA expected the nation's unconventional gas production to continue soaring for decades. (Shown here is production for the contiguous United States, excluding Alaska.)

Hubbert had been largely right about the peak and decline of conventional gas. But he hadn't thought there was much potential for unconventional gas—and had been proven wrong about that.

The shale gas boom was so prolific that supply rose faster than demand. Gas prices crashed, from over $8 per million cubic feet in 2008 to under $3 in 2012. This made natural gas suddenly much less attractive, and drillers searched for the next boom. It turned out the same techniques also worked for tapping oil, unleashing another boom that became known as shale oil, or more accurately as tight oil.*

THE US OIL boom came at just the right time. The world's conventional oil production reached a peak in 2006, according to International Energy Agency statistics, just over 70 million barrels per day, and then declined marginally over the following years.

The IEA had issued reassurances before that this kind of peak wouldn't occur in the foreseeable future. But after doing in-depth analyses of thousands of the world's oil fields, the IEA had come to a much more pessimistic view. In its 2010 *World Energy Outlook*, it expected conventional oil "never regains its all-time peak of 70 million barrels per day reached in 2006." It foresaw conventional production remaining on a plateau, almost perfectly flat, through 2035.

Nonetheless unconventional sources such as tar sands and tight oil could keep the total supply of oil rising through the 2030s, the IEA expected. The catch was that all the alternatives had high costs of production. So Fatih Birol, the IEA's chief economist, declared, "The age of cheap oil is over."

The tight oil boom became even more critical after 2011, when pro-

---

* Shale oil is not to be confused with oil shale, which has to be cooked to release oil. Because of this confusion, many prefer the term *tight oil*. Also *tight oil* is more geologically accurate because in the most prolific plays—like the Bakken in Montana and North Dakota, and Texas's Eagle Ford—companies weren't actually fracking shale; they were targeting more porous layers of rock that were interspersed with shale, as in a tiramisu.

tests erupted in the Middle East and North Africa, a movement dubbed the "Arab spring." In Libya an uprising toppled the longtime dictator, Muammar Gaddafi, and as the nation descended into civil war, its oil production plummeted. Civil war likewise broke out in Syria, and its production dwindled. Meanwhile the United States was enforcing sanctions against Iran to try to prevent it from developing nuclear bombs, and with its export possibilities limited, Iran lowered its oil production. By 2014, more than 2 million barrels a day of production—3 percent of the world's total—had been wiped out.

But the US boom filled the gap. "It's really just a coincidence, but they have perfectly matched, canceling each other out," remarked Christof Rühl, BP's chief economist, in June 2014. As a result, oil prices had remained remarkably stable, around $100 a barrel, for a few years. It was, as Rühl put it, a period of "eerie calm."

The US oil boom continued accelerating. But at the same time, world economic growth was lackluster, still suffering a hangover from the Great Recession. Given these conditions, the world had a limited appetite for barrels carrying a $100 price tag. Something had to give.

The price of oil began to slide in the summer of 2014. Many analysts expected OPEC to react by cutting production to prevent a worldwide glut and to keep oil prices up. Instead, OPEC decided to continue producing at the same rate—and as a result, oil prices fell further, eventually dropping by half by the end of the year and settling around $50 a barrel through 2015.

In response, America's tight oil drillers cut back their activity drastically. Still, through mid-2015, they continued adding enough new wells to hold production roughly steady. The tight oil boom was heralded as surprisingly resilient.

Meanwhile, however, shale drillers' finances were not so resilient. Many companies carried such a heavy debt load that the interest ate up a large chunk of their earnings. There was growing concern about the financial health of shale drillers, and whether they would manage to repay their loans and avoid bankruptcy. The Bank of International Settlements—the Switzerland-based adviser to central banks around the world—analyzed these drillers, concluding, "Highly leveraged producers may attempt to maintain, or even increase, output levels even as the oil price falls in order to remain liquid and to meet interest payments and tighter credit conditions." What looked like resilience could actually be desperation.

An analysis by Bloomberg Intelligence of sixty-two shale drillers found that before the price crash, they'd been spending twice as much as their revenues—and after the crash, this cash burn became even larger, with spending four times as high as revenues. As a result those drillers took on more debt, rising by the spring of 2015 to $235 billion. "The question is," said an analyst at ratings agency Standard & Poor's, "how long do they have that they can get away with this?" David Einhorn, a

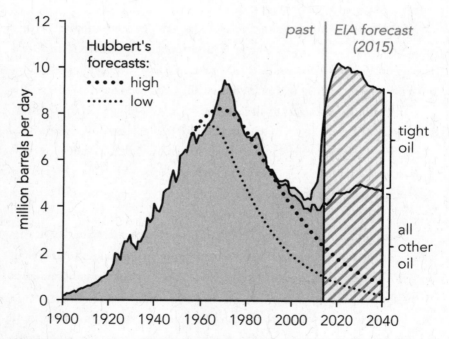

US oil production forecasts: Hubbert's oil production forecasts were prescient, with conventional production declining in line with his expectation. In 2015 the EIA expected the nation's tight oil to peak by 2020 but to continue yielding a significant amount for decades to come. It also expected a long-term revival of other sources of oil. (Shown here is production for the contiguous United States, excluding Alaska.)

famed hedge fund manager, made waves in May 2015 when he declared shale drillers' business models a failure. "None of them generated excess cash flow, even when oil was at $100 a barrel," Einhorn said. "Aside from a few choice locations, they don't earn a positive return on capital."

Nonetheless, this tight oil boom had managed to put a bump on the Hubbert curve—and what a bump it was.

The Energy Information Administration expected the tight oil boom to continue growing into the 2020s, then to peak and only gradually taper off over decades. It also expected a revival of conventional oil production, so that the nation's total production would break the former all-time high, even if only briefly.

But would this great revival fizzle sooner, dropping off more sharply?

OVER THE YEARS, both the US Energy Information Administration and the International Energy Agency repeatedly overestimated how much conventional oil would materialize. Meanwhile these agencies—along with most everyone else—repeatedly underestimated how quickly the US fracking boom would unfold and repeatedly revised their long-term forecasts for shale gas and tight oil. As an IEA analyst remarked about the US boom, "We keep raising our forecasts, and we keep underestimating production."

So in years to come, could such forecasts keep going up and up?

There are two very different reasons that past forecasts could have come out too low, at least in the short run. They may have underestimated the long-term scale of the boom—and thus low-balled how much it would ultimately supply. Or they may have underestimated the pace of the boom—in which case, for a given size of resource, higher production in the early years means the resource is simply getting used up more quickly than expected.

It's tricky to gauge how long the shale boom might last. While various shale formations can span large areas, most of the production to date comes from relatively small areas where wells are more prolific—known as "sweet spots." Through the early years of the boom, the industry and researchers alike were still figuring out how big those sweet spots were and where they faded into less attractive rock. Another facet, likewise uncertain, was how tightly wells could be packed into the sweet spots. If drilled too close together, they would interfere with each other, so each well wouldn't yield enough oil or gas to turn a profit.

In the face of such complexity, the EIA had used assumptions somewhat like the old Zapp hypothesis. In its modeling of shale plays, EIA assumed that drillers could sink wells spaced evenly across a whole region, their horizontal legs all lined up like the rows of a cornfield. They also assumed each future well across a whole play would be at least as productive as those already drilled. By 2013 the EIA had refined that approach, applying that assumption county by county. This gave a somewhat more detailed view—but it was still fairly coarse-grained, since US counties often cover some five hundred square miles, each large enough to contain thousands of shale wells.

The resolution of the assessment matters because the sweet spots are relatively small portions of each play, surrounded by large areas that yield far less gas or oil. Companies naturally try to target the sweet spots first— so after the sweet spots are tapped out, the remaining rock is generally lower quality, meaning future wells in those areas may be less productive. So the EIA's method—which assumed future wells would be at least as productive as past wells—could easily lead to overly optimistic results.

To do a more detailed analysis of America's major shale gas plays, a team at the University of Texas, Austin, undertook a series of in-depth studies starting in 2010. The team divided each shale play into square-mile blocks, looking at the performance of wells in each block, to clearly distinguish sweet spots from marginal areas. By examining how quickly production from each well dropped off, they were able to estimate how much rock each well drained. In this way, the team estimated how tightly wells could be packed in before they'd significantly interfere with each other. By mapping the locations of all existing wells, and simulating the addition of new wells only where there was enough space to slot them in among their neighbors, the team got a realistic sense of how many wells could actually fit into each play. The team also excluded drilling under major cities such as Pittsburgh, and under protected lands such as national forests.

The Texas team's detailed modeling was arguably the most rigorous shale gas analysis published to date—and its resulting forecasts were significantly lower than EIA's and lower still than most industry forecasts. For America's "big four" shale gas plays—Barnett, Fayetteville, Haynesville, and Marcellus—the Texas team forecast the collective production would peak by 2020 and then enter a sharp decline.

This forecast was based on the EIA's expectation that natural gas prices would gradually rise, from under $4 per million cubic feet in 2014 to about $7.50 by 2040 (in inflation adjusted dollars). Given that same price scenario, the EIA expected production from those "big four" plays to soar significantly higher through the late 2020s, then to plateau at least until 2040. The difference was stark.

All such forecasts have large uncertainty, and both the Texas team and the EIA explored a range of possibilities. However, to get results similar to the EIA's, the Texas team had to crank up all its assumptions to levels that were "too optimistic," it concluded.

The results were "bad news," said Tad Patzek, then head of the University of Texas at Austin's department of petroleum and geosystems engineering, and a member of the team conducting the analyses. (Patzek was also a longtime admirer of Hubbert, having worked at Shell Oil's Bellaire lab—albeit long after Hubbert left.) With efforts to extract shale gas as fast as possible and to export significant quantities, the nation was working on the assumption that shale gas would be abundant for decades to come. But in light of the Texas team's results, Patzek argued, "We're setting ourselves up for a major fiasco."

If the Texas team's studies turn out to be on the mark, the shale gas boom could peter out surprisingly soon. As of late 2015, the team was also studying two major tight oil plays, the Bakken and the Eagle Ford, but had yet to publish the results. If their tight oil forecasts likewise come out lower than many others' analyses, then the United States—and the world—could be heading for an oil fiasco as well.

. . .

CONVENTIONAL CRUDE IS the foundation of world oil markets—and yet the prospects for increased production seem slim. After peaking in 2006 at 70 million barrels per day, according to IEA statistics, conventional oil production declined slightly, to under 69 million barrels in 2014.

Just to hold conventional production flat, compensating for the decline of older fields, the IEA estimates that each year the industry needs to add about 4 million barrels a day of new production—that's like adding another China, the world's fourth-largest oil producer, every year. The drop in conventional oil production, even through a decade of rising oil prices, suggests that the oil industry simply hasn't been able to find and develop enough new fields to outpace declines. As Tim Dodson, CEO of Norwegian oil giant Statoil, said in 2015, "The industry is struggling big-time to replace their oil resources and reserves."

In 2010, the IEA's *World Energy Outlook* forecast that conventional production would likely remain on a plateau for decades. In the subsequent years, IEA lowered its expectations for conventional oil, forecasting a decline. Others have subsequently followed a similar pattern. In 2015, when BP released its latest long-term energy forecasts, it showed conventional production on a plateau through 2035. Exxon's 2015 outlook likewise had conventional oil essentially flat through 2040.

It brings to mind the old saying about the adoption of a challenging idea: First people ignore it, then they deny it, then they say it was obvious all along. Although not widely talked about, a consensus formed that conventional oil production would go no higher than its level in the mid-2000s—so that could turn out to be the all-time peak. Such forecasts have been continually revised downward over the past decade, as conventional oil repeatedly fell short of expectations—so it seems likely these forecasts could continue to be revised downward in coming years. If so, that would mean the coming decline will be sharper than most have been expecting.

Following the conventional oil peak in 2006, total oil production nonetheless continued rising. All the growth was due to unconventional sources, such as US tight oil and Canada's tar sands. These sources have risen rapidly, but they still supply less than 10 percent of the world's oil.

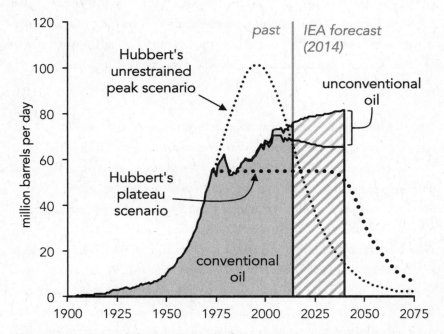

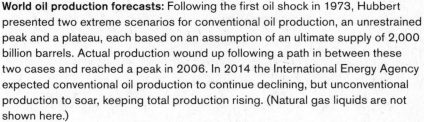

World oil production forecasts: Following the first oil shock in 1973, Hubbert presented two extreme scenarios for conventional oil production, an unrestrained peak and a plateau, each based on an assumption of an ultimate supply of 2,000 billion barrels. Actual production wound up following a path in between these two cases and reached a peak in 2006. In 2014 the International Energy Agency expected conventional oil production to continue declining, but unconventional production to soar, keeping total production rising. (Natural gas liquids are not shown here.)

They're all relatively expensive to extract, requiring large investments. With the oil price drop of 2014, it was unclear how many unconventional projects would still be profitable at an oil price of, say, $50 a barrel. Prices dropped so low that even many conventional oil projects looked unprofitable. Within a year of the start of that price drop, oil companies delayed or canceled over $200 billion of projects. Another $1.5 trillion of investment was at risk.

These cutbacks in investment could lead to lower production than had been expected—and could set up another price spike in a few years. The industry would welcome such higher prices, as they may be necessary to enable companies to keep production rising. But rising prices could also take a toll on the global economy. So the world may be caught in a trap, in which people are unwilling or unable to continue paying for oil at prices of $100 or more, but production isn't able to continue rising, either.

OVER THE PAST several years, there's been increasing acceptance of the notion of peak oil—albeit in a different guise.

"There is plenty of oil in the world, there really is," Tony Hayward, chief executive of oil giant BP, argued in 2010. He expected that "world demand will peak before its supply peaks," and that this turning point would "probably occur beyond 2020."

Others thought it might be much sooner. In 2013 the *Financial Times* argued that world oil seemed to be headed "towards self-imposed peak production and a well managed shift" to other energy sources. Later, *The Economist* joined the chorus, arguing, "We believe that oil is close to a peak. This is not the 'peak oil' widely discussed several years ago. We believe that demand, not supply, could decline." Even as oil remained the world's number-one source of energy, *The Economist* declared it "yesterday's fuel." In 2015, Daniel Yergin agreed, "There will be worldwide peak oil demand. We think it's going to be in the 2030s."

These were strange arguments coming from oil companies, financial publications, and veteran analysts, since they spoke of "supply" and "demand" while saying little about the price of oil. However, demand is always relative to the price—that's Economics 101. At record-high oil prices, people would surely consume less—and if oil prices were to fall to $5 a barrel, it's easy to imagine that people would find ways to consume much more. ("Who wants to fly to Las Vegas this weekend?") With high oil prices, the world may soon hit peak demand—but there's little sign that it is approaching peak desire.

For decades, Americans had been driving steadily more and more miles each year, but this trend reached a peak in 2005 at 10,000 miles per person, then fell over the next couple of years—and fell further amid the Great Recession. A new phrase came into being: *peak car*. Finally, perhaps, prices at the pump had gotten high enough that Americans were driving less.

However, after oil prices dropped to around $50 a barrel by the end of 2014, Americans quickly began driving more. Sales of gas-guzzlers increased—not just in the United States but also in China. In the summer of 2015, a vice president at General Motors said, "We just wrapped up the US auto industry's best six months in a decade."

"Peak demand" and "peak supply" are just two sides of the same— well, the same barrel. It seems peak demand may not come unless it's forced upon consumers, one way or another—by high prices at the pump, or by recessions, or even by strict climate regulations.

IF CONVENTIONAL OIL production has peaked, what can we do?

Faced with the prospect of shortages, the dominant approach around the world has been to wait for market signals—that is, higher prices—to spur change. But market signals are often volatile, driven by short-term gluts or shortages rather than reflecting a long-term outlook.

So far the main effect of higher oil prices has been to spur unconventional oil projects. Their long-term potential, in particular for tight oil, is highly uncertain. If sources such as tight oil don't last as long as many hope, and other sources of energy or ways of running the economy aren't scaled up in time, the world could find itself in a bind.

Market forces alone seem unlikely to achieve a smooth transition away from oil over the long run. If conventional oil production were to remain roughly flat for decades to come—as the IEA, BP, and Exxon expect—then coping could call for long-term planning, especially if the world is going to cut greenhouse gas emissions at the same time. Or if conventional oil production were to decline more sharply, as Hubbert's work suggests it could, that would almost certainly call for some form of planning to help societies adapt.

It's possible that there really will be plenty of unconventional oil, of various forms, for many decades to come. However, burning all that additional fuel would only further increase greenhouse gas emissions, which need to be curtailed quickly if the world is to limit global warming. Coping with the peak of conventional oil while also tackling climate change means the options are more limited. But there are synergies, in that many strategies would help tackle both problems at the same time.

To make such historic energy transitions, Hubbert certainly supported long-term planning, with massive government projects on energy research and development as just a starting point. Throughout his life, he was skeptical about the widespread reliance on markets and their price signals, and questioned the profit motive as a driving force in society. So far, through the age of oil, when the fuel was cheap, people consumed it wantonly, without worrying much about how future generations would get by. Companies naturally invested in what was profitable in the near-term. Government spending and policies on alternatives—such as nuclear, wind, and solar—have been variable and relatively small-scale, compared with the challenge of transforming the world's energy systems.

Instead of devoting efforts to a transition away from oil, nations around the world have focused on maintaining economic growth in the short run. Growth remains as much of a sacred cow as it was throughout Hubbert's life. He believed that giving up the obsession with growth was the truly daunting challenge.

HUBBERT'S OIL FORECASTS weren't correct in every detail. He was, of course, not actually an oracle. But they were strikingly prescient.

However, Hubbert's predictions have often been misunderstood and misrepresented. Commentators have claimed he'd said that production—for any particular nation or region—must follow a bell-shaped curve. That he didn't know about, or ignored, unconventional oil such as tar sands. That he likewise ignored reserve growth. That he completely disregarded prices, or assumed that technologies would remain static.

However, Hubbert did address these issues in his studies. He was, for example, the first to develop a systematic way of quantifying reserve growth for all oil fields in the United States. His focus was on conventional oil, but he did state estimates for unconventional oil. And the long-term decline of conventional oil production in the United States, and the start of a similar decline worldwide, have matched Hubbert's expectations closely—strikingly so, considering that his forecasts are now several decades old.

While Hubbert's forecasts became well known, his wider ideas about society did not. The term *technocrat* has entered the popular lexicon, but the group Technocracy has been largely forgotten. His worries about hitting limits to growth, and his calls for slowing growth before the limits are harshly forced upon humanity, though little talked about, still haunt us today. (The whole problem of climate change is but one example of a limit to growth—that is, a limit to the greenhouse gases we can spew into the atmosphere before we fundamentally change the character of our planet.)

Hubbert also posed deep questions about modern life. How are goods allocated between citizens? How do we deal with inequality and poverty? Can democracies plan for the long term? And finally, what constitutes a good life?

Having grown up hardscrabble in central Texas, Hubbert appreciated how technology transformed our ways of living, and he saw fossil fuels as a boon that humanity should not squander. These fuels drove the great flourishing of science, which Hubbert saw as "the only thing standing between us and the Dark Ages."

Hubbert spent much of his life studying and debating limits to fossil fuels. Because of this he was widely branded a pessimist. But he was in many ways a great optimist. As he put it in his early twenties, "I like people and have a great deal of faith in them, in spite of all failings." Although he was a curmudgeon and a nitpicker, he retained this faith in people throughout his life. He never ceased to hope that with education and discussion, people could take an approach to planning for the future that was far-sighted, rational, and—perhaps most important—fair.

Although he celebrated modern technology, he didn't see material wealth as an end in itself. In 1976, as the United States struggled with stagflation, unemployment, and other aftereffects of the first oil shock, he told a journalist:

> We're entering into something new. It could be a cataclysm, but it doesn't have to be. It need not even be an energy crisis. The transition from exponential growth over to a stabilized state, or nongrowth, need not be a state of intellectual decay. In fact, it could easily be a renaissance—an intellectual renaissance, a golden age. Because with our technology and with adequate supplies of energy, we ought to have a lot of leisure. And the proper use of this leisure can bring us an intellectual renaissance.
>
> The cultural adjustments that must be made, and can be made, could easily lead to a flowering of civilization whereby we would look at the mental state we're in now as the Dark Ages, culturally.

Hubbert was, throughout his life, a curious mixture. He was a hard-nosed scientist who crafted rigorous studies, but he also risked his reputation with bold forecasts. He was impatient with what he saw as irrational thinking, but he was eternally hopeful that humanity would reach a greater maturity, with a better understanding of its position on the planet and its place in the long span of time. He could sometimes be a doomsayer but was more often a dreamer.

SOURCES

INTERVIEWS BY THE AUTHOR

Bartlett, Albert. October 22, 2011, by phone
Bethke, Craig. September 7, 2013, and May 6, 2015, by phone
Broussard, Martha Lou. June 18, 2012, by phone
Cordesmeyer, Paul. November 3, 2012, at Technocracy's Continental Headquarters in
    Ferndale, WA
Deffeyes, Kenneth. February 14, 2012, by phone
Downey, Marlan. May 15, 2012, by phone
Hall, Charles. May 2, 2013, by phone
Hirsch, Robert. June 21, 2012, by phone
Hubbert, Michael. February 14, 2013, in Boonville, CA
Laherrère, Jean. April 27, 2011, in Brussels, Belgium
Meadows, Dennis. November 4, 2011, at the Cosmos Club, Washington, DC
Miller, Betty. July 29, 2012, by phone
Pollock, David. June 12, 2012, by phone
Priest, Tyler. May 1, 2013, by phone
Rose, Peter. May 10, 2013, by phone
Schlesinger, James. August 17, 2012, by phone
Skinner, Brian. April 4, 2011, by phone
Stegemeier, George. July 31, 2013, by phone

COLLECTIONS USED, WITH ABBREVIATIONS:

AHC: M. King Hubbert papers, American Heritage Center, University of Wyoming
Bretz: J Harlen Bretz papers, University of Chicago
Chaplin: Ralph Chaplin papers, University of Michigan
Chase: Stuart Chase papers, Library of Congress

Davis-AHC: Morgan J. Davis papers, American Heritage Center, University of Wyoming

Davis-UT: Morgan J. Davis papers, University of Texas, Austin

Hopkins: Harry Hopkins papers, Franklin D. Roosevelt Presidential Library

Marx: Guido Marx papers, Stanford University

McKelvey: Vincent E. McKelvey papers, American Heritage Center, University of Wyoming

NAS: National Academy of Sciences and National Research Council archives, Washington, DC

Pinchot: Gifford Pinchot papers, Library of Congress

Rubey: William W. Rubey papers, Library of Congress

Rowe: James Rowe papers, Franklin D. Roosevelt Presidential Library

SZ: M. King Hubbert letters in possession of Suzie Zupan, Hubbert's grandniece

TIP: Technocracy, Inc., papers, at its Continental Headquarters in Ferndale, WA

(Where collections have numbers and names for the boxes and folders, these are abbreviated in the endnotes. So box 12, folder 3, for example, is written as B12 F3.)

ABBREVIATIONS (FOR ENDNOTES AND BIBLIOGRAPHY)

AAPG: American Association of Petroleum Geologists

BEW: Board of Economic Warfare

CSC: Civil Service Commission

CT: *Chicago Tribune*

DMN: *Dallas Morning News*

Doel: Hubbert, Interviews by Ronald Doel

EIA: US Energy Information Administration

IEA: International Energy Agency

LAT: *Los Angeles Times*

MKH: M. King Hubbert

NAS: National Academy of Sciences

NRC: National Research Council

NYHT: *New York Herald-Tribune*

NYT: *New York Times*

OGJ: *Oil and Gas Journal*

TSC: *Technocracy Study Course*

WP: *Washington Post*

WSJ: *Wall Street Journal*

NOTES

PROLOGUE

xii   **"Couldn't you tone"**: Doel pt. 7.

xii   **"The age of"**: Inman, "Has the World."

xv   **"twentieth-century geology's"**: "M. King Hubbert at 100: The Enduring Contributions of Twentieth-Century Geology's Renaissance Man," session at the annual meeting of the GSA, Seattle, WA, November 2003.

xv   **"This outspoken maverick"**: US NRC, Committee on Opportunities in the Hydrologic Sciences, *Opportunities in the Hydrologic Sciences* (Washington, DC: National Academies Press, 1991), p. 72.

CHAPTER 1: THE JOURNEY

3   **Chicago's defining feature:** This description of the city draws from Bessie Louise Pierce and Joe Lester Norris, *As Others See Chicago* (Chicago: University of Chicago Press, 1933), and from Simon Baatz, *For the Thrill of It* (New York: Harper, 2008).

3   **"Lessen the smoke":** "The Tribune's Platform for Chicago," *CT*, February 13, 1919.

4   **"I remember wishing":** The quotes in this and the next seven paragraphs are from Doel pt. 1.

5   **"the Harvard of the West":** Quoted in Alonzo Ketcham Parker, "The First Year: October 1, 1892, to October 1, 1893—Continued," *University Record*, January 1, 1917, p. 52.

5   **"That was my first":** Doel pt. 1.

6   **"If you can't make":** Quoted in Studs Terkel, *American Dreams, Lost and Found* (New York: Pantheon, 1980), p. 88.

6   **Hubbert . . . was overwhelmed:** MKH, *Technocracy Study Course*, p. 192.

6   **"when they looked":** Doel pt. 1.

7   **"nearly killed me":** Ibid.

8   **"I'd learned that":** Ibid.

9   **"Was it real":** Doel pt. 2.

9   **"You haven't declared":** Doel pt. 1.

10   **J Harlen Bretz:** Soennichsen, *Bretz's Flood*; Doel pt. 2.

10   **"I haven't committed":** Doel pt. 1.

10   **"about $60 between":** MKH to Nell, July 27, 1927 (SZ).

10   **"Not one college fellow":** Ibid.

CHAPTER 2: PIONEERS' EYES

11   **"Not a thing":** Doel pt. 2.

11   **"That was a godsend":** Doel pt. 1.

12   **"What are you looking " and "It was as if":** Doel pt. 2.

12   **"steeped in cynicism":** The quotes in this and the next paragraph are from MKH to Nell, September 19, 1925 (SZ).

13   **"We've found out":** Quotes in this paragraph and the next are from Doel pt. 2; also see MKH, "Bretz's Baraboo."

CHAPTER 3: BUGGED OUT

14   **"Now, here's this credit":** The dialogue in this and the next paragraph is from Doel pt. 2.

15   **"Before with literature":** Ibid.

16   **"The students of":** Ibid.

16   **"one of the major":** Ibid.

16   **"We learned that" and "get our hands dirty":** MKH, "Bretz's Baraboo."

16   **"My God it was":** Doel pt. 2.

16   **"impossible . . . miners would never":** "Report of the Committee on the Nomenclature of Faults," *Bulletin of the GSA*, January 1913, p. 175.

17   **"I reviewed it" and "Why don't we":** Doel pt. 2.

17   **"M. King Hubbert, University":** MKH, "Suggestion for Simplification."

18   **"I didn't think":** The quotes in this and the next four paragraphs are from Doel pt. 2.

19   **"Within 50 years":** MKH, Geology 102 notebook, spring 1926 (AHC, B69 F "Geology 102").

19   **"Very young industry":** Ibid.

19   **"That course was":** Doel pt. 2.

CHAPTER 4: COMPLETE REBELLION

20   **"New Oil Town":** "New Oil Town Is a Modern Sodom," *Richardson* (TX) *Echo*, June 11, 1926. For the description of Borger, see Olien and Olien, *Life in Oil Fields*, p. 40.

20   **"the biggest, wildest":** MKH, Interview by Andrews.

20   **"No one knows":** Quoted in Hinton and Olien, *Oil in Texas*, p. 139.

21   **"by far the" and "an inexhaustible supply":** "Suggested West Texas Drive," *DMN*, May 2, 1926.

21   **"on top of":** "Panhandle Rig Building Difficult," *DMN*, March 13, 1926.

21   **Railroad Commission:** Hinton and Olien, *Oil in Texas*, chap. 6.

22   **"like so many buzzards":** MKH, *Technocracy Study Course*, p. 170.

22   **"got into a" and "almost ill feeling":** Doel pt. 2.

23   **"Why didn't you" and "Well, I'm finished":** Doel pt. 3.

23   **"see if they could":** MKH to Nell, November 21, 1926 (SZ).

23   **"Well, it was":** Doel pt. 3.

23   **"I've developed an awful":** MKH to Nell, November 21, 1926 (SZ).

24   **"I needed a little":** Doel pt. 2.

25   **"We only had one":** Ibid.

26   **"It would have meant":** Quotes in this paragraph are from Doel pt. 2.

26   **"I haven't looked":** MKH to Nell, May 20, 1928 (SZ).

27   **"quite a series":** Doel pt. 2.

27   **"I began to stew":** Ibid.

27   **"when you can measure":** Quoted in MKH, "Is Being Quantitative Sufficient?"

28   **"Geophysics is not" and "Usually their notions":** Doel pt. 3.

28   **"I was entirely":** Doel pt. 2.

28   **"complete rebellion over" and "ringleader":** Ibid.

28   **"I am opening up":** MKH to Nell, March 2, 1929 (SZ).

29   **"doesn't mean that":** Quoted in Christopher Cerf and Victor S. Navasky, *The Experts Speak* (New York: Pantheon, 1984), p. 57.

29   **"Financial Storm Definitely":** Quoted in James Grant, *Bernard M. Baruch* (New York: John Wiley & Sons, 1997), p. 227.

30   **"All this has happened":** MKH to Nell, April 14, 1930 (SZ).

30   **"I was a general neophyte":** Doel pt. 2.

30   **"I have had about":** MKH to Nell, October 25, 1930 (SZ).

30   **"I used to think":** MKH to Nell, August 6, 1930 (SZ).

30   **"It is really terrible":** MKH to Nell, March 9, 1930 (SZ).

30   **"Have busted with Cora":** MKH to Nell, August 6, 1930 (SZ).

31   **"dragged me into":** MKH to Nell, January 10, 1931 (SZ).

31   **"It seems that":** MKH to Nell, August 6, 1930 (SZ).

CHAPTER 5: THE MEETING

35   **"I have blundered":** MKH to Nell, February 26, 1931 (SZ).

36   **"I was so fascinated":** Ibid.

36   **"A criminal is":** Quoted in MKH, *Technocracy Study Course*, p. 176.

36   **technical director of Muscle Shoals:** MKH to Nell, February 26, 1931 (SZ).

36   **Theory of Energy Determinants:** For Scott's ideas, see Ackerman, "Origin of Technocracy," and "The Facts Behind Technocracy," *Scholastic*, February 18, 1933

(part 1), March 4, 1933 (part 2), and March 18, 1933 (part 3); Akin, *Technocracy*; BEW, "Questioning of M. King Hubbert"; Chase, *Technocracy*; CSC, "Report" and "Inserts"; Doel pt. 4; Elsner, *Technocrats*; John S. Gambs, *The Decline of the IWW* (New York: Russell & Russell, 1966); MKH, *Technocracy Study Course*; Bassett Jones, "Production vs Consumption," private memo, 1932 (Pinchot papers, B2554 F "Unemployment—Technocracy [A], AS 1932"); Loeb, *Life in a Technocracy*; Raymond, December 1932 series in *NYHT* and *What is Technocracy?*; Scott, "Scourge," "Political Schemes," "Technocracy Speaks," and "Technology Smashes"; Scott et al., *Introduction to Technocracy*; Walsh, "Engineers' Revolution."

36   **"We have before us"**: Scott, "Technology Smashes."

39   **"genius"**: MKH to Nell, July 24, 1932 (SZ); also see Doel pt. 4.

39   **the Technical Alliance**: Larry Anderson, *Benton MacKaye* (Baltimore: Johns Hopkins University Press, 2008), p. 136; Anon., "Technical Alliance—early MSS," c. 1920 (TIP); Joseph Dorfman, *Thorstein Veblen and His America* (New York: Viking Press, 1934), pp. 460–62; Paul Harrison, "Technocracy Is Enemy of Waste," Newspaper Enterprise Association in *Daily News* (Frederick, MD), December 28, 1932; Bassett Jones to Michael Angelo Heilperin, September 16, 1932 (Pinchot papers, B2554 F "Unemployment—Technocracy [A], AS 1932"); Guido Marx to Joseph Dorfman, May 10, 1934 (Marx papers, B1 F6); "The Industrial Engineer," editorial, *New York Call*, March 19, 1920.

39   **chief engineer**: Technical Alliance brochure, c. 1920 (TIP); Ackerman, "Origin of Technocracy."

41   **"didn't have any more"**: Doel pt. 4.

42   **"It seems to me"**: MKH to Nell, February 26, 1931 (SZ).

42   **"Woman Without Country"**: "Woman Without Country Seeking Citizenship," *Appleton* (Wisconsin) *Post-Crescent*, September 12, 1931.

43   **"But why did you"**: Quoted in Kenneth Norman Stewart, *News Is What We Make It* (New York: Houghton Mifflin, 1943), p. 104.

44   **"It's the newest thing"**: Walsh, "Engineers' Revolution."

44   **"Perhaps you would"**: MKH to Nell, July 24, 1932 (SZ).

CHAPTER 6: THE FAD

46   **"the first step"**: "Toward a New System," *Nation*, September 7, 1932.

46   **"We are now"**: "Technocrats," *Time*, September 26, 1932.

46   **"What I wanted"**: Doel pt. 4.

47   **"a general avidity"**: Columbia University, *Annual Report for the Year Ending June 30, 1933* (New York: Columbia University, 1933), p. 364.

47   **Roosevelt's longer-term outlook**: Collins, *More: Politics of Growth*, and Kennedy, *Freedom from Fear*.

47   **"It seems to me"**: Franklin D. Roosevelt, "Address at Oglethorpe University in Atlanta, Georgia," May 22, 1932.

47   **"our last frontier"**: Franklin D. Roosevelt, "Campaign Address on Progressive Government at the Commonwealth Club in San Francisco, California," September 23, 1932.

48   **"I do challenge"**: Herbert Hoover, "Address at Madison Square Garden in New York City," October 31, 1932.

48   **"It is not necessary"**: "Technocracy," *WSJ*, November 18, 1932.

48   **"Is the machine"**: "The Machine Age," AP in *Reading* (PA) *Eagle*, October 24, 1932; "Will a Machine Get Your Job?" *LAT*, November 6, 1932.

49   **"They are wrong" and "They overlook the fact"**: "Motor Official Criticizes Theories of Technocracy," *WP*, December 26, 1932.

49   **"This Technocracy thing"**: Will Rogers, "Mr. Rogers Is Still Puzzled by This Technocracy Thing," *NYT*, December 23, 1932.

49   **"What do you think"**: Ellis, *Nation in Torment*, p. 222.

49   **"Technocracy? *Was Das*?" and "*Ja. . . . The problem*"**: "Technocracy? Wass Dass? Asks Einstein," AP in *Morning Herald* (Gloversville and Johnstown, NY), January 10, 1933.

49   **"decreased the need"**: "Einstein Says Slump Causes Are Internal," AP in *Morning Herald* (Gloversville and Johnstown, NY), January 24, 1933.

50   **"Technocracy Will Not Work"**: "Technocracy Will Not Work, Benito Mussolini Declares," *San Antonio* (TX) *Light*, February 2, 1933.

50   **too many fantastical stories:** Howard Scott federal personnel file, US National Archives and Records Administration; Raymond, "Prophet of Technocracy Surrounded by Legends," *NYHT*, December 17, 1932; Raymond, *What Is Technocracy?* pp. 102–11; "Technocracy's Question," *New Outlook*, December 1932; "Technocrat," *Time*, December 26, 1932; "Technocrat Scott Known in Oregon," *Oregonian*, January 15, 1933; Perkins, "Technocracy Leader"; "Scott as Polish Mixer Was Curiosity, Now Pompton Lakes Recalls Virtues," *New York World-Telegram*, January 21, 1933; "Chief Technocrat Was a Wax Maker," *NYT*, January 22, 1933.

50   **"Dr. Scott"**: This and the quotes in the next two paragraphs are from Raymond, *What Is Technocracy?* pp. 109–10.

51   **"The real Scott"**: Ibid., p. 100. Scott had also claimed, in applying for his job at Muscle Shoals, to have degrees from three German universities, at Freiburg, Leipzig, and Munich. Howard Scott federal personnel file, US National Archives and Records Administration.

51   **"There is not"**: Rogers, "Mr. Rogers Is Puzzled."

51   **"Everybody from bank"**: "Technocracy! It's a Modern Word—For What?" *CT*, December 25, 1932.

51   **"At first we"**: "Shop Talk at Thirty," *Editor & Publisher*, December 17, 1932.

52   **"We are not attempting"**: "400 Leaders Hear Technocracy Plea," *NYT*, January 14, 1933.

52   **"then I am afraid"**: "Won't Seek Technocrat Dictator, Scott Says, but Politicians May," *New York World-Telegram*, January 14, 1933.

52 **"The whole country"**: Harry Elmer Barnes, "Mr. Scott and Technocracy," *New York World-Telegram*, February 2, 1933.

52 **"Of course it" and "That decided it"**: Elsner, *Technocrats*, p. 14; also see "Technocracy Ranks Split," *Columbia Spectator,* January 24, 1933.

53 **"Scott Is Ousted"**: "Scott Is Ousted from Technocracy by Split in Group," *NYT,* January 24, 1933; "Rebels Oust Technocracy High Priest," *Oakland Tribune,* January 24, 1933.

53 **"Everybody I talk to"**: "Recovery Is Here, Asserts Mr. Ford," *WSJ,* February 2, 1933. See also "Times Good, Not Bad, Ford Says," *NYT,* February 1, 1933.

53 **"Reds and possible Communists"**: Quoted in Ellis, *Nation in Torment*, p. 268.

53 **"Let me assert my firm belief"**: Quoted in ibid., pp. 268–72.

54 **"So much hell"**: MKH to Nell, May 18, 1933 (SZ).

54 **"You had better"**: Ibid.

CHAPTER 7: LESSONS

55 **"monad"**: Adamson and Moore, *Technocracy: Questions Answered*, p. 25.

55 **"comprehensive treatise"**: MKH to Chaplin, October 17, 1933 (Chaplin papers, B1 F "Hubbert, Marion King").

55 **"I am going"**: MKH to Chaplin, August 8, 1933, ibid.

55 **"irrefutable and loaded" and "I want to put"**: MKH to Chaplin, October 17, 1933, ibid.

56 **"I've got to write"**: MKH to Nell, January 16, 1934 (SZ).

56 *Technocracy Study Course*: The book does not state an author, but in the Library of Congress copyright listings, the various lessons are credited to M. King Hubbert, with the exception of the final one, Lesson 22, which is credited to Technocracy, Inc. The writing throughout has a similar style and uses phrases that Hubbert used through much of his life, suggesting he wrote the whole *TSC* (with the exception of the "Introduction," which was credited to Arch Jamieson, another Technocracy member). In 2013 current members of Technocracy told me that their understanding was that Hubbert had written all the lessons. Page numbers listed here refer to the 1940 edition.

56 **"this was in the deep dark"**: Doel pt. 5.

57 **"It has come"**: *TSC*, p. 99.

58 **"life history of" and "The production rises"**: Ibid., p. 102.

58 **"the mad business"**: Ibid., p. 170.

59 **"The persistent S-shape"**: Ibid., p. 98.

59 **"In the United States"**: Ibid., p. 104.

59 **"The rapidity with which" and "Where does Japan"**: Technocracy memorandum, "Subject: Study Course," December 15, 1934 (TIP).

59 **"Two years ago"**: "Technocracy Gains, Doctor Hubbert Says," *Democrat-Chronicle* (Rochester, NY), December 31, 1934.

60   **"I saw those boys"**: Doel pt. 5

61   **"We had to have"**: This and remaining quotes in the paragraph are from MKH, testimony to CSC, "Report."

61   **"The whole visit"**: MKH to Rubey, February 4, 1935 (Rubey papers, B5 F8).

61   **"The period of"**: *TSC*, p. 167.

61   **"The American Price system"**: "Scientific Prophecy," *Technocracy* ser. A, no. 3 (August 1935), p. 19.

61   **"would in short order"**: This and the following quotes are from MKH, "Some Facts of Life."

62   **"No question of"**: *TSC*, p. 177.

63   **a network of garages:** Ibid., p. 254.

63   **"regulation dress"**: Akin, *Technocracy*, p. 194; Anon., notes on *Time* magazine interview with Howard Scott, May 11, 1940 (TIP).

63   **lapel pins:** Pettijohn, *Memoirs*, p. 163; Alfonso A. Narvaez, "M. King Hubbert, 86, Geologist Who Influenced Oil Production," *NYT*, October 17, 1989.

CHAPTER 8: BORDERLANDS

65   **"I'll take that"**: Doel pt. 3.

65   **As a child:** The mouse, windmills, and puddles are recalled ibid.; MKH, *Structural Geology*, p. 7.

65   **"It would be impossible"**: Galileo Galilei, *Dialogues Concerning Two New Sciences*, trans. Henry Crew and Alfonso de Salvio (New York: Macmillan, 1914), pp. 130–31.

66   **"What material did"**: MKH, *Structural Geology*, p. 8; Doel pt. 3.

68   **"hard rock" and "soft earth" theories:** MKH, "Strength of the Earth."

68   **"the paradox of an"**: MKH, "Theory of Scale Models."

69   **"quite a ripple"**: MKH to Nell, January 10, 1938 (SZ).

69   **"Dr. Hubbert."**: "Report on Second Meeting of the Interdivisional Committee on Borderland Fields . . . ," December 4–5, 1936, p. 3 (NAS, NRC Central Files, F "G&G: Com on Borderland Fields between Geology Physics & Chemistry; Interdivisional").

CHAPTER 9: THE SWIFTEST DECLINE

70   **"I have only now"**: MKH to Nell, January 10, 1938 (SZ).

70   **"Otherwise it would be"**: MKH, testimony to CSC, "Report."

70   **"We have politely"**: MKH to Nell, January 10, 1938 (SZ).

71   **Henderson aired his concerns:** Brinkley, *End of Reform*, pp. 23–24; "Leon Henderson Rules Our Economic Roost," *CT*, October 5, 1941.

71   **"the major part of this"**: Laughlin Currie, "Causes of the Recession" (Hopkins papers, Federal Relief Agency papers, B55 F "Lauchlin Currie report April 1, 1938").

71   **"the swiftest decline"**: "Recessional," *Time*, November 22, 1937.

71   **"As far back as we"**: MKH, "Determining the Most."

72   **"the period of exponential" and "While we do not"**: MKH, "Determining the Most."

73   **"We're going to put"**: MKH to Nell, January 10, 1938 (SZ).

73   **he drew 2,500 in Akron**: "Howard Scott Tour Striking Success!" *Technocracy* ser. A, no. 15 (December 1938), p. 20.

74   **"Am considering getting"**: MKH to Nell, March 3, 1939 (SZ).

CHAPTER 10: FIGHTING MAD

77   **"My results are"**: MKH to William Rubey, February 5, 1940 (Rubey papers, B5 F8).

77   **"I relaxed for"**: Doel pt. 3.

78   **"an embarrassing blunder"**: MKH to Rollin Chamberlin, March 27, 1940 (AHC, B178 F "Theory of Groundwater Motion [4]").

78   **"a sizable corps"**: MKH to Chamberlin, November 18, 1940, ibid.

78   **"My experience at Columbia"**: Quotes in this paragraph and the next are from MKH to Eckart, May 27, 1940, ibid.

79   **"synthesize it into"**: MKH to Eckart, May 27, 1940, ibid.

79   **"hibernate . . . all apple carts"**: MKH to Chamberlin, June 24, 1940, ibid.

79   **"I didn't feel it was"**: MKH, testimony to CSC, "Report."

79   **"emotional and abusive" and "fighting mad"**: Doel pt. 3.

79   **"if you are wrong"**: BEW, "Questioning of M. King Hubbert."

80   **"Where does Japan"**: Technocracy memorandum, "Subject: Study Course," December 15, 1934 (TIP).

80   **"it didn't pan out"**: MKH, testimony to CSC, "Report."

CHAPTER 11: TOTAL MOBILIZATION

81   **"many false alarms"**: MKH to William Rubey, February 5, 1940 (Rubey papers, B5 F8).

81   **a long list of contacts**: MKH, "Record S 780-R," 1941–1942, black notebook (AHC, B81, loose).

82   **"the New Deal's"**: "The Week in the Nation," *Social Justice*, December 5, 1938, p. 9.

82   **"price czar"**: "Leon Henderson Rules Our Economic Roost," *CT*, October 5, 1941.

82   **"Leon Henderson has"**: "Henderson, Aides Accused by Dies," *NYT*, September 8, 1941.

82   **"eat on the Treasury"**: "Henderson Link to Reds Charged," *LAT*, September 8, 1941.

82   **"one of the craziest"**: "Dies Says M'Leish Hires Communist," *NYT*, January 16, 1942.

82   **"the loudest, fightingest"**: "What Price Henderson," *Collier's*, September 6, 1941.

83 **"My offer still"**: "Dies Says M'Leish Hires Communist," *NYT*, January 16, 1942.

83 **"Director-General of Defense"**: Technocracy CHQ, "Howard Scott for Director-General of Defense," December 31, 1941 (TIP); also quoted in Elsner, *Technocrats*, p. 154.

83 **full-page advertisements:** The earliest ad I could find that called for Scott to be director-general of defense was in *Seattle Post-Intelligencer*, January 20, 1942. Subsequent similar ads were in *Oregonian* and *San Diego Union* (January 21), *LAT* (February 3), and *Cleveland Plain Dealer* (February 10). The ads without mention of the director-general position were in *NYT* (March 8) and *WP* (March 11).

83 **"This is Continental Headquarters"**: "Technocracy's Back Clicking Heels and Saluting," *New York Post*, March 17, 1942.

84 **"crashing the gate"**: MKH to Rubey, February 5, 1942 (Rubey papers, B5 F8).

84 **"at least 35 high"**: "Nudism Book Studied," AP in *CT*, March 30, 1942.

84 **"The high technological"**: Maurice Parmelee, *Farewell to Poverty* (New York: John Wiley & Sons, 1935), p. 472.

85 **"life sans clothing"**: "Delights of Life Sans Clothing Read Into Congressional Record," *WP*, March 31, 1942.

85 **"If I have ever seen"**: "Dies Attacks Stir Debate in House," *NYT*, March 31, 1942.

85 **"reorganization"**: "Parmelee Demands BEW Reinstate Him," *NYT*, April 20, 1942.

85 **"Strategic Minerals from"**: MKH, BEW notebook, July 6, 1942 (AHC, B104 F "Board of Economic Warfare . . .").

85 **"science and engineering"**: Quotes in this and the next paragraph are from MKH, testimony to CSC, "Report."

86 **"There has not been"**: MKH to Scott, January 10, 1943 (TIP).

86 **"irresponsible, unrepresentative, crackpot"**: "Dies Lists 40 on US Rolls as Crackpots," *CT*, February 2, 1943.

87 **"we may as well"**: Harold Loeb, *Production for Use* (New York: Basic Books, 1936), p. 97.

87 **"I know of"**: "Names and Records of 40 Crackpots," *CT*, February 2, 1943.

87 **"rated ineligible" and "requested to separate"**: Moyer to BEW chairman, April 3, 1943 (TIP).

87 **"I have no"**: This and other dialogue in this chapter is from BEW, transcript, April 14, 1943 (TIP).

88 **compiling a list:** Lists of groups deemed subversive, or alleged to be so, are in James H. Rowe papers (B41 F "Subversive Activities"). Technocracy was not on Dies's lists, nor was it listed as subversive in the files of the House Committee on Un-American Activities (HUAC), based on: Rowe papers (B42 F "Subversive Activities"); Rep. John S. Wood to Leland A. Shankland, March 21, 1950, with attached report from HUAC on Technocracy, Inc. dated March 20, 1950 (TIP); "Technocracy, Inc.," HUAC report for Senator Henry M. Jackson, December 21, 1953 (TIP).

90 **"pyramiding of the"**: MKH, "Economic Transition."

92 **"The dominant figure"**: Quotes in this paragraph and the next are from Lawson Moyer to BEW, May 31, 1943 (TIP).

93 **"next to the President"**: Quoted in Steven Fenberg, *Unprecedented Power: Jesse Jones, Capitalism, and the Common Good* (College Station: Texas A&M University Press, 2011), p. 368.

93 **"Well, it was insured"**: Doel pt. 4. The story has also been told other places, such as John K. Galbraith, *Name-Dropping: From F.D.R. On* (Boston: Houghton Mifflin, 1999), p. 39. However, the story may not be true. In 1943, after Stuart Chase wrote about it in *The Nation*, Jesse Jones wrote a letter of complaint to Chase, saying he never made that comment, and Chase apologized, saying he'd read it in another article in *The Nation* (Chase to Jones, April 3, 1943, in Chase papers, B14 F1). In any case, Hubbert and others told the story for decades afterward since it epitomized a common outlook.

94 **"obstructing the war effort"**: "Feud Breaks Out Over War Buying," AP in *Lawrence* (KS) *Journal-World*, June 29, 1943.

94 **"They seem to have"**: MKH to Scott, July 16, 1943 (TIP).

94 **"a new school"**: "Ballet Dancer Takes a Whirl at News," *CT*, August 1, 1943.

94 **"disgusted"**: Doel pt. 4.

CHAPTER 12: NO DOUBT

97 **"a very interesting"**: Doel pt. 4.

97 **"keep in touch"**: Doel pt. 5.

98 **"There was a bit"**: Ibid.

98 **"Fortunately they had" and "work like hell"**: Ibid.

99 **"the city the Depression"**: Quoted in Joe R. Feagin, *The New Urban Paradigm* (Lanham, MD: Rowman & Littlefield, 1989), p. 160.

99 **"The only place" and "a hell of an"**: Doel pt. 5.

99 **Catholics disliked that the center**: Maria Helen Anderson, "Private Choices vs. Public Voices: The History of Planned Parenthood in Houston," Rice University doctoral thesis, 1998, chap. 4.

100 **"debut"**: Doel pt. 5.

100 **"had no smell" and "in a weak position"**: Ibid.

100 **"Will the rock" and "The eyebolts would"**: MKH, "Strength of the Earth."

102 **"I was all right"**: Ibid.

CHAPTER 13: THE MILLION-DOLLAR LAB

103 **Shell Oil had fallen behind**: On Shell in the 1940s, see Priest, *Offshore Imperative*, chap. 2.

104 **"The only drawback"**: MKH to Howard Scott, June 15, 1945 (TIP).

104 **"since almost all" and "No expense will be":** MKH to William Rubey, September 4, 1945 (AHC, B10 F2).

104 **"there will be no water-tight":** MKH to Marshall Kay, November 28, 1945 (AHC, B19 F3).

105 **"on somebody else's":** Doel pt. 5.

106 **masterful lecturer:** Fred M. Meissner, "M. King Hubbert as a Teacher," presented at the annual meeting of the GSA, Seattle, WA, November 2003.

106 **"What became apparent":** Doel pt. 5.

107 **"I learned this":** Ibid.

CHAPTER 14: A PRECARIOUS POSITION

108 **"amazed governmental and" and "American motorists have":** "Shortages of Gasoline, Fuel Loom Despite Record Output," *NYT*, February 6, 1948.

108 **"It was beginning to look":** "The US Runs Short of Oil," *Life*, February 9, 1948.

109 **"far too much time":** "Synthetic Fuel Output Proposed to Congress," *LAT*, January 27, 1948.

109 **"The production center now":** "Krug Urges US Aid to Build Synthetic Fuel Industry," *WSJ*, February 10, 1948.

110 **"Our senses have":** MKH, "Energy from Fossil Fuels," draft (AHC, B133 F3); see also "Solar Radiation Figured Sufficient to Support Man," *Christian Science Monitor*, September 25, 1948.

110 **Populations around the world:** See Joel E. Cohen, "Population Growth and Earth's Human Carrying Capacity," *Science* 269, no. 5222 (July 21, 1995), pp. 341–46, doi:10.1126/science.7618100.

113 **"use our rich heritage":** "Atom Power for Space Ships to Defy Gravity Envisioned," *NYT*, September 16, 1948.

113 **"That was the first":** Doel pt. 6

CHAPTER 15: THE PARLEY

114 **"Good God almighty":** Doel pt. 6.

116 **"periodical nervous attacks" and "I am sure":** United Nations, *Proceedings of Scientific Conference*.

116 **"an exercise in metaphysics":** "Estimate of 500-Year Oil Supply Draws Criticism in U.N. Parley," *NYT*, August 23, 1949.

116 **"perfectly objective estimates" and "If it continues":** "Excerpt of Transcript on United Nations Scientific Conference," August 22, 1949 (AHC, B179 F "1949 UN Meeting").

117 **"no one knows":** United Nations, *Proceedings*.

117 **"Oil companies usually":** "More Oil Than We Think," *NYT*, August 26, 1949.

117   **"decidedly embarrassed":** Doel pt. 6.

118   **"The philosophy that" and "nobody knows how":** "Earth's Supply of Oil Unknown, Scientists Say," AP in *CT*, August 27, 1949.

118   **"If I hadn't done it":** Doel pt. 6.

119   **"Once this peak":** MKH, "Report on Aspects of UN Conference."

120   **"Scott did give a date":** Cordesmeyer interview by author.

120   **"Technocracy Inc. has been":** Norwin Kerr Johnson, "Technocracy Scores Again," *Technocrat*, June 1938. Scott's predictions were reported in many newspapers and magazines, such as Walsh, "Engineer's Revolution"; "400 Leaders Hear Technocracy Plea," *NYT*, January 14, 1933; Perkins, "Technocracy Leader"; "Predictions by Technocracy Are Borne Out, Scott States," *NYT*, May 14, 1933; "Capitalism Due for Fall, Scott states," *Spokane* (WA) *Press*, October 8, 1937; "Insists Technocracy Is 'Cure,'" AP in *NYT*, April 26, 1938.

120   **"money economy . . . goods economy":** "Economic Warfare Troubles," *Chemical and Metallurgical Engineering*, April 1942.

121   **"That glass bead pack":** Deffeyes interview by author.

121   **"That Hubbert is a bastard":** Quoted in Deffeyes, *Hubbert's Peak*, pp. 3–4. Other sources: Deffeyes interview by author; Stegemeier interview by author; Wilson, "Hubbert's Curve."

CHAPTER 16: STAMPEDE

122   **"The theory was valid" and "grab off an oil field":** Doel pt. 5; see also MKH, "Prospecting for Hydrodynamic."

122   **"I wrote a very hot":** Doel pt. 5.

123   **"not only are these effects":** MKH, "Entrapment of Petroleum."

123   **"There was a stampede":** Doel pt. 5.

124   **"must have set the":** Marlan Downey, "Recalling Geochem, Fluid Flow Advances," *AAPG Explorer*, October 2003.

124   **"potentially one of":** MKH to H. R. Aldrich, December 6, 1951 (AHC, B18 F6).

124   **"geologist's geologist":** "Wallace E. Pratt Memorial Grant-in-Aid," AAPG website, accessed November 27, 2010.

124   **"It explains many facts":** Pratt to MKH, October 31, 1953 (AHC, B13 F13).

124   **"spectacular results":** "Presentation of Day Medal to M. King Hubbert," *Geological Society of America Proceedings for 1954* (July 1955), pp. 65–68.

124   **"a classic":** A. I. Levorsen, *Geology of Petroleum* (San Francisco: W. H. Freeman, 1954), p. 563.

125   **"It has been purely fortuitous":** MKH, "Prospecting for Hydrodynamic," p. 19; see also MKH, "Entrapment."

CHAPTER 17: A MAGICAL EFFECT

126 **"got a dogfight going"**: Doel pt. 6.

126 **"fracking"**: "'Fracking'—A New Exploratory Tool," *OGJ*, October 12, 1953.

128 **"Due to availability"**: J. B. Clark, "A Hydraulic Process for Increasing the Productivity of Wells," *Petroleum Transactions, American Institute of Mining and Metallurgical Engineers*, January 1949, pp. 1–8, doi:10.2118/949001-G.

129 **"accepted almost universally"**: This and the remaining quotes in this chapter are from Doel pt. 6.

CHAPTER 18: SWEEPING UNDER THE RUG

133 **"We were just about everything"**: The quotes in this and the next four paragraphs are from Doel pt. 5.

134 **"several restless nights" and "neither dignified"**: MKH to Gilluly, April 28, 1955 (AHC, B19 F3).

134 **"I seem to have an aptitude"**: MKH to A. J. Galloway, April 28, 1955 (AHC, B19 F3); MKH to H. S. M. Burns, April 28, 1955 (AHC, B19 F3).

134 **"As you well know"**: MKH to Gilluly, April 28, 1955 (AHC, B19 F3).

135 **"I was very skeptical"**: Doel pt. 5.

135 **"confidential" or "secret"**: Ralph E. Lapp, "Nuclear Power Secrecy," *Bulletin of the Atomic Scientists* 12, no. 4 (April 1956), p. 135.

135 **"To some extent"**: Quotes in this and the next four paragraphs are from NRC, *Disposal of Radioactive Waste*.

136 **"A little bit of uranium"**: Doel pt. 5.

137 **"Dr. M. King Hubbert prefers"**: "Minutes of the Meeting of the Steering Committee, Conferences on the Disposal of Radioactive Waste Products in Geologic Structures," NAS-NRC, September 12, 1955, p. 2 (NAS, Division of Earth Sciences, Committee on GARWD—Meetings).

CHAPTER 19: JOLTED

138 **"for the foreseeable future"**: This phrase was regularly used by the petroleum industry and by journalists reporting on it; examples include "World Oil Supply Held Biggest Ever," AP in *NYT*, February 28, 1952.

138 **"a critical look"**: MKH to William Strang, December 14, 1955 (AHC, B121 F "American Petroleum Institute").

139 **"a reappraisal of"**: MKH to Harold Gershinowitz, February 8, 1956 (AHC, B180 F "Presentation 'Energy Resources of the World'").

139 **tweaked these numbers**: MKH, "Nuclear Energy and the Fossil Fuels," preprint. Hubbert's figures—and the numbers I use throughout the book, unless otherwise noted—refer only to crude oil and condensate. Others would sometimes give

numbers for "oil" or "petroleum liquids" that also include natural gas liquids, or sometimes even include biofuels. For definitions of these various liquid fuels, see the EIA's online glossary, http://eia.gov/tools/glossary/index.cfm.

140 **"Production of oil":** American Petroleum Institute press release, March 8, 1956 (AHC, B156 F "Nuclear Energy and the Fossil Fuels [2]").

141 **"That part about reaching the peak":** Doel pt. 7.

141 **"several continents, a number":** This and the following quotes from the talk are from MKH, "Nuclear Energy and the Fossil Fuels," preprint. Press accounts of his talk used exactly the same wording as his paper.

145 **"End of Oil Boom":** For this and other media coverage, see "End of Oil Boom After 10 More Years Forecast," AP in *Corpus Christi Times*, March 9, 1956; "Oil Crisis Seen Close," *San Antonio Express*, March 9, 1956.

146 **"drastically in error":** MKH, "Nuclear Energy and Fossil Fuels," preprint.

146 **"It jolted the hell":** Magida, "He Predicted Oil Shortage."

146 **"They just about":** MKH, Interview by Andrews.

146 **"I didn't introduce":** Magida, "He Predicted Oil Shortage."

CHAPTER 20: GUESSTIMATING

149 **"I wish it had been":** Wallace Pratt to MKH, March 13, 1956 (AHC, unknown folder).

149 **"The paper is meaty":** Pratt to MKH, March 28, 1956 (AHC, B13 F13).

150 **"This dependence is sure":** US Geological Survey, "The Oil Supply of the United States," *Bulletin of the AAPG* 6, no. 1 (1922), pp. 42–46, doi:10.1306/3d9325c5-16b1-11d7-8645000102c1865d. The estimate of remaining oil was 9 billion barrels, including both reserves (oil "in sight") as well as more speculative "prospective and possible" oil fields.

150 **"I just couldn't":** Pratt to Hubbert, March 28, 1956 (AHC, B13 F13).

151 **"conservative" and "the industry feels":** Pratt, "Impact of Atomic Energy," pp. 90, 94.

151 **Humble had submitted a forecast:** Humble Oil report for Pratt and cover letters dated December 1, 1955 (Davis-UT papers, B3X302 F2).

151 **"has reacted to King":** Pratt to Davis, March 29, 1956 (Davis-UT papers, unknown box).

151 **"all the rest of us":** Pratt to Ray Walters, September 6, 1977 (AHC, B50 F13).

151 **"was perfectly furious":** Ibid.

152 **"The whole thing":** Doel pt. 7.

152 **"a marble mausoleum":** MKH, Interview by Andrews.

152 **"a certain amount":** Doel pt. 7.

152 **the most pessimistic geologist:** W. C. Lowrey to MKH, January 10, 1974 (AHC, B133 F1); see also Doel pt. 7.

154 **Morgan Davis:** "Humble's Indispensable Derrick: Morgan Davis," *Alcalde*

(November 1960); Wallace E. Pratt and Dean A. McGee, "Morgan J. Davis," *AAPG Bulletin* 65 (September 1981), pp. 1950–52.

154 **"in conflict with":** Quotes in this and the next two paragraphs are from Morgan Davis, "Exploration Outlook," May 17, 1956 (Davis-UT papers, BSX324 F6).

155 **"Capital, energy, and freedom":** Richard J. Gonzalez, "What Makes America Great?" speech delivered May 9, 1951; republished on the website of the American Heritage Education Foundation.

155 **"The dynamic petroleum":** Richard J. Gonzalez, "We Are Not Running Out of Oil," *Dealers' Digest*, November 19, 1956. The rhetoric that Davis and Gonzalez used is an early example of a technique, later common, of spreading doubt as a way of supporting the status quo. Corporations and think tanks would employ it to counter evidence that smoking causes cancer and, later, evidence that burning fossil fuels causes global warming. See Oreskes and Conway, *Merchants of Doubt*.

156 **"The thinking that you":** William Bradley to MKH, October 8, 1956 (AHC, B133 F1).

156 **Pratt was known:** Everette DeGolyer, "Wallace Everette Pratt: First Sidney Powers Memorial Medalist: An Appreciation," *Bulletin of the AAPG* 29, no. 5 (May 1945), pp. 47890.

156 **"I am a little embarrassed":** Pratt to MKH, October 12, 1956 (AHC, B13 F13).

156 **Pogue was curious:** Joseph Pogue to MKH, December 9, 1956 (AHC, B156 F "Nuclear Energy and the Fossil Fuels [2]")

157 **"best possible estimate":** Magida, "He Predicted Oil Shortage."

157 **"Methods of Predicting":** Quotes in this and next four paragraphs are from MKH, "Methods of Predicting . . ." notebook entry, December 10, 1956, pp. 319–23 (AHC, B144 F "Geophysical Notes for Vol VI [2]").

159 **"the advantage of":** Quotes in this and next three paragraphs are from MKH to Joseph Pogue, December 26, 1956 (AHC, B156 F "Nuclear Energy and Fossil Fuels [2]").

CHAPTER 21: THE BEER-CAN EXPERIMENT

160 **"tremendous pressure beneath":** B. Mostofi and August Gansser, "The Story Behind the 5-Alborz," *OGJ*, January 21, 1957, pp. 78–84.

160 **"very significant bit":** MKH to William Rubey, February 6, 1957 (Rubey papers, B5 F8).

160 **a detour through the Swiss Alps:** Doel pt. 6; MKH and Rubey, "Role of Fluid Pressure"; MKH, *Structural Geology*, pp. 1, 15–16.

161 **"would put me in an asylum":** Quoted in Richard Fortey, *Earth: An Intimate History* (New York: Knopf, 2004), p. 97.

161 **"out pops a":** MKH to William H. Freeman, October 4, 1957 (AHC, B18 F1).

163 **"The idea is so":** MKH to Rubey, October 1, 1957 (Rubey papers, B5 F8).

CHAPTER 22: PRESTO! CHANGO!

164 **"the greatest public works"**: "Highway Project History's Greatest," *Milwaukee Sentinel*, May 30, 1956.

164 **Suez Canal:** Yergin, *Prize*, chap. 24.

165 **"all you had to do"**: Bennett H. Wall, *Growth in a Changing Environment* (New York: McGraw-Hill, 1988), p. 579.

165 **"quite wild"**: The quotes in this paragraph are from MKH to Sam Schurr, January 15, 1957 (AHC, B10 F9); Hubbert later published a critique of Netschert's book: MKH, "Review of *The Future Supply*."

165 **16 million barrels a day:** Netschert, *Future Supply*, p. v.

165 **Chase Manhattan:** Kenneth Hill et al., "Future Growth of the World Petroleum Industry," *Proceedings, American Petroleum Institute*, Section IV—Production (1957), pp. 105–17.

165 **"generally accepted figure"**: "Drillers Can Look to a Bright Future," *OGJ*, October 21, 1957.

166 **"Are we likely"**: Quoted in MKH to Wallace Pratt, January 28, 1958 (AHC, B13 F13).

166 **"persistent propaganda line"**: Quotes in this and the next paragraph are in ibid.

166 **"completely invalidated"**: Thomas Nolan, "The Inexhaustible Resource of Technology," in Henry Jarrett, ed., *Essays on America's Natural Resources* (Baltimore: Johns Hopkins University Press, 1958). Also published as Thomas Nolan, "Use and Renewal of Natural Resources," *Science* 128, no. 3325 (September 19, 1958), pp. 631–36, doi:10.1126/science.128.3325.631.

167 **"The prediction of"**: MKH to Nolan, April 24, 1958 (AHC, B10 F9).

168 **"on the threshold of"**: Quotes in this and the next two paragraphs are from MKH, "Mineral Resources of Texas."

169 **"it was not exactly what"**: MKH to Bretz, October 20, 1958 (AHC, B14 F6).

169 **"Throughout all this time"**: Doel pt. 7.

170 **"the hazards of"**: Quotes in this and the next two paragraphs are from Morgan Davis, "The Dynamics of Domestic Petroleum Resources," November 12, 1958 (Davis-AHC papers, B1 F9).

170 **"Bright Picture Painted for Oil"**: "Bright Picture Painted for Oil," *NYT*, November 13, 1958.

170 **"remarkably like one"**: MKH to William Rubey, August 22, 1958 (Rubey papers, B5 F8).

171 **"If you never see"**: MKH to Rubey, February 25, 1959 (AHC, B4 F5).

171 **"I couldn't really believe"**: Doel pt. 7.

172 **"And yet that curve"**: Ibid.

172 **"This business about"**: Quoted in Yergin, *Prize*, pp. 538.

172 **"the certified requirements"**: Eisenhower, "Statement by the President Upon Signing Proclamation Governing Petroleum Imports," March 10, 1959.

173 **"At least one controversy":** "3,000 Expected at Geology Meet," AP in *Big Spring* (TX) *Herald*, March 15, 1959.

173 **"There is sound reason":** "AAPG Meet Talks Differ on New Discovery Views," *DMN*, March 17, 1959.

173 **"I think we can rule":** MKH, "Techniques of Prediction with Application."

174 **"a rat race":** Quoted in MKH to Robert B. McNee, September 5, 1962 (AHC, B23 F11). See also Doel pt. 7; MKH, "Discussion by G. Moses Knebel," memo, April 28, 1959 (AHC, B112 F "Prediction Paper—Notes [2]").

174 **"There are powerful voices":** MKH to Philip Abelson, April 21, 1959 (AHC, B181 F "1959-04-27 to 29 . . .")

174 **"wouldn't stay still":** Quotes in this paragraph are from Drew, *Undiscovered Petroleum*, p. 40.

174 **"meaningless":** Quotes in this and the next paragraph are from MKH to Preston Cloud, June 5, 1959 (AHC, B156 F "Nuclear Energy and the Fossil Fuels [2]").

175 **"I pulled the paper":** Drew, *Undiscovered Petroleum*, p. 40.

CHAPTER 23: THE RICH TEXANS

176 **"as much as 600 billion":** Quoted in Kaufman, *Project Plowshare*, p. 40.

177 **"There is enough oil":** Edward Teller, "Nuclear Miracles Will Make Us Rich," *This Week*, February 1, 1959.

177 **"irresponsible and almost":** MKH, draft letter on Teller and Project Plowshare, June 16, 1956 (AHC, B19 F9).

177 **"Project Screw-ball":** MKH to Louis McCabe, January 28, 1959, ibid.

177 **two hundred cases of leukemia:** Edward Teller and Albert L. Latter, *Our Nuclear Future: Facts, Dangers, and Opportunities* (New York: Criterion, 1958), p. 119.

177 **"carrying out a determined":** MKH, draft letter on Teller and Project Plowshare, June 16, 1956 (AHC, B19 F9).

178 **"fundamentally opposed":** MKH, "Visit of Dr. Gerald Johnson," memo, June 16, 1959 (AHC, B4 F6).

178 **"the growth bug":** "The Growth Bug," *CT*, September 7, 1960.

178 **"a rising tide":** Quoted in Ted Sorensen, *Counselor: A Life at the Edge of History* (New York: HarperCollins, 2008).

179 **"intellectual ferment":** MKH, "Telephone Call from Robert C. Cook, Population Reference Bureau," memo, December 1, 1960 (AHC, B21 F3).

179 **"How much of this":** Doel pt. 7.

179 **"vast areas of":** E. B. Germany, "The Economy of Texas," in Herbert Gambrell, ed., *Texas Today and Tomorrow* (Dallas: Southern Methodist University Press, 1961).

179 **"replete with reassuring":** Quotes in this and the following paragraph are from MKH, "Discussion of Paper by E. B. Germany: 'The Economy of Texas,'" December 1960 (AHC, B181 F "Philosophical Society of Texas").

180 **"especially spirited"**: Philosophical Society of Texas, *Proceedings of the Annual Meeting*, p. 4.

180 **"That so-and-so!"**: Doel pt. 7.

180 **"I thoroughly enjoyed"**: MKH to George McGhee, December 12, 1960 (AHC, B181 F "Philosophical Society of Texas").

180 **"pretty contemptuous of"**: Doel pt. 7.

CHAPTER 24: A GRAND SURVEY

181 **"We need your help"**: John F. Kennedy, "Remarks Before the National Academy of Sciences," April 25, 1961.

181 **"grand survey"**: "A Grand Survey of Natural Resources," *New Scientist*, August 31, 1961.

181 **"extremely sophisticated questions"**: Kennedy, "Remarks Before."

182 **"our entire society"**: NRC, *Natural Resources: Summary Report*, p. 1.

182 **"This impression may"**: MKH to Detlev Bronk, May 1, 1961 (AHC, B156 F17).

182 **"we are headed"**: MKH to Edward Espenshade, May 1, 1961, ibid.

183 **"Horseshit!"**: This quote and the quotes in the next eight paragraphs are from Doel pt. 7.

185 **"near exhaustive exploration,"** Zapp, "Future Petroleum," p. H-23.

185 **"seems little doubt"**: Zapp, "World Petroleum," p. 2.

185 **"like plowing a cornfield"**: MKH to A. A. Meyerhoff, c. March 1965, draft (AHC, B105 F "Energy Resources—Major [2]"). See also MKH, "Is Being Quantitative."

185 **"Those who are"**: Quoted in MKH, *Energy Resources: Report to Committee*, p. 48.

185 **"Look, that 300 billion"**: This quote and dialogue in next two paragraphs is from Doel pt. 7.

186 **"trash"**: Doel 7.

186 **590 billion barrels**: McKelvey to MKH, July 20, 1962 (AHC, B153 F "NAS Committee on Natural Resources—Gen Corr [1]").

186 **"I didn't have any"**: Doel pt. 7.

186 **AAPG's statistics**: MKH to McKelvey, August 17, 1962 (AHC, B153 F "McKelvey corr").

186 **"Look, I can't even"**: Doel pt. 7.

186 **"mind was made up"**: Ibid.

187 **"Attempting to assess"**: US Senate, Committee on Interior and Insular Affairs, *National Fuels and Energy*, pp. 68–70.

187 **"there is plenty"**: Ibid., p. 12.

188 **"about the worst offender"**: MKH, "Memorandum on conference with Wallace Thompson," June 14, 1962 (AHC, B134 F "Data [subsequent] notebook, 1962, excerpts").

188 **"that asshole in"**: Doel pt. 8.

188 **"Well, gentlemen"**: Quotes in this and the next two paragraphs from ibid.

189 **"the peak of production"**: MKH, *Energy Resources: Report to Committee*, p. 63.

189 **Pratt had estimated:** Wallace Pratt, "Oil Fields in the Arctic," *Harper's*, January 1944.

189 **"Every field which"**: MKH, *Energy Resources: Report to Committee*, p. 67.

191 **"contingency allowance"**: Ibid., pp. 72–73.

191 **"That's all doctored"**: The dialogue here and in the next six paragraphs is from Doel pt. 7.

192 **resign in protest:** Ibid. See also NAS-NRC Central File, section "Committees & Boards," Folder "Com on Natural Resources—General," in particular Edwin R. Gilliland to Detlev Bronk, August 3 and 23, 1962.

192 **"Gentlemen, what do you"**: Doel pt. 7.

193 **"Most oil-company sponsored"**: MKH to Robert B. McNee, September 5, 1962 (AHC, B23 F11).

193 **"an unbiased compendium"**: "Fuel Study is Praised by INGAA," *Houston Post*, September 23, 1962.

193 **"sensible, fair, and objective" and "limited to facts"**: "Oil, Gas Score High in Senate Energy Report," *Houston Post*, September 23, 1962.

193 **"We've Found Half"**: "We've Found Half Our Oil, Says Hubbert," *OGJ*, January 21, 1963.

193 **"I fully expect"**: MKH to Philip Abelson, February 15, 1963 (AHC, B133 F11).

193 **"In the midst"**: "Amount of Oil in Ground Is Not Current Problem, Says Oilman," *Pampa* (TX) *Daily News*, February 10, 1963.

194 **"If Hubbert's forecast"**: "Reserves Guess Far Wrong," *Houston Post*, February 18, 1963.

194 **"None of the managers"**: Kenneth Deffeyes to MKH, February 12, 1963 (AHC, B34 F2).

194 **"Not the least"**: Wallace Pratt to MKH, February 5, 1963 (AHC, B133 F11).

194 **"There was a little"**: US Congress, Joint Committee on Atomic Energy, *Development, Growth, and State*, pt. 1, p. 168.

197 **"the Nation's fossil-fuel reserves"**: Ibid., pt. 1, p. 92.

197 **"known reserves minable"**: Ibid., pt. 1, p. 97.

197 **"My analyses lead"**: Ibid., pt. 2, p. 895.

CHAPTER 25: PENNY PINCHING

198 **"they can't do this"**: US Congress, Joint Committee on Atomic Energy, *Development, Growth, and State*, pt. 1, p. 182.

199 **"hoping that they"**: Doel pt. 5.

199 **"should in no sense"**: NRC, *Disposal of Radioactive Waste*, pp. 3–4.

200 **"didn't like the criticism"**: Doel pt. 5.

200 **"We set up our own"**: Congress, *Development, Growth, and State*, pt. 1, pp. 179–84.

201 **"he sounded kind of"**: Doel pt. 8.

201 **"She wanted me"**: Ibid.

201 **"penny pinching"**: MKH, "Telephone Call from John Higgins, Business Week," memo, June 7, 1973 (AHC, B60 F2).

CHAPTER 26: RETURNING HOME

202 **"years of servitude"**: MKH to Harlen Bretz, January 7, 1955 (AHC, B14 F6).

202 **"I was perfectly happy"**: Doel pt. 8.

202 **"As you know"**: Quotes in this paragraph are from Thomas Nolan to MKH, November 6 and December 12, 1963 (AHC, B48 F18).

202 **"I had no objection"**: Doel pt. 8.

203 **"the lead dog"**: Deffeyes interview by author.

203 **"What am I"**: Pratt to MKH, September 17, 1963 (AHC, B34 F2).

204 **reduction in drilling costs:** Downey interview by author; MKH to Rubey, June 6, 1966 (AHC, B169 F "Rubey, William"); "How Shell Detects High-pressure Formations on the Gulf Coast," *OGJ*, May 16, 1966.

204 **"sending money to"**: Deffeyes interview by author.

204 **"Hubbert is a bastard"**: Quoted in Deffeyes, *Hubbert's Peak*, pp. 3–4.

204 **"He didn't care"**: Quotes in this and the next paragraph are from Broussard interview by author.

205 **"national influence"**: Doel pt. 8.

205 **"I was trying" and "the influence of"**: Magida, "USGS Plays."

205 **"I feel as if"**: MKH to Nolan, December 2, 1963 (AHC, B32 F5).

205 **"I should have"**: Doel pt. 8.

205 **"virtually set his own"**: "Texan Appointed to Federal Post," *DMN*, February 12, 1964.

205 **"I found out that there'd"**: Doel pt. 8.

206 **"it would be most"**: William Pecora to MKH, February 2, 1965 (McKelvey papers, B4 F "Hubbert").

206 **"I am equally concerned"**: MKH to A. A. Meyerhoff, unsent draft, c. March 1965 (AHC, B105 F "Energy Resources—Major [2]").

206 **"one of the more serious"**: MKH to Nolan, March 4, 1965 (AHC, B105 F "Ryan discussion, petroleum statistics [1]").

CHAPTER 27: THE ZAPP HYPOTHESIS

208 **Hendricks's effort:** Hendricks, *Resources of Oil, Gas*.

208 **"obtained by pure assumption"**: MKH, "Commentary on Manuscript by Bernardo F. Grossling," January 26, 1976 (AHC, B42 F9).

208 **"working at saturation"**: MKH to E. F. Osborn, May 3, 1965 (AHC, B27 F6).

208 **"hectic, if not"**: MKH to Nell Jessup, June 12, 1963 (SZ).

208 **"When I first"**: Doel pt. 8.

210 **No one . . . had done this:** J. R. Arrington at Carter Oil had also estimated future reserve growth and worked with some of the same data sets as Hubbert. But he hadn't published a systematic method for forecasting future reserve growth.

212 **"the present analysis":** Quotes in this and the next paragraph are from MKH, "Degree of Advancement."

212 **"I'll fix you!":** Quotes in this and the next two paragraphs are from Doel pt. 8.

213 **"The logic was":** Doel pt. 9.

213 **an old story:** Stillman Drake, *Galileo at Work* (Chicago: University of Chicago Press, 1978), pp. 19–21, 414. Some authors have surmised that the Pisa experiment is an apocryphal story, but Drake argues that the evidence suggests it was real.

213 **"Well, that's a pretty":** Doel pt. 9.

214 **the State Department:** MKH, "Telephone Call, Preston E. Cloud, Jr.," memo, June 18, 1968 (AHC, B60 F1).

215 **"no significant changes":** MKH, "Telephone Call from Pres. Cloud," memo, June 14, 1968 (AHC, B163 F "Resources & Man, notes—environment—Energy Resources—Biosphere").

216 **Akins warned:** Stobaugh and Yergin, *Energy Future*, pp. 3, 267; Yergin, *Prize*, p. 568.

216 **"was a triggering":** Quoted in Yergin, *Prize*, pp. 598–99.

CHAPTER 28: RUTHLESS

217 **"an inevitable age":** "Public Fights A-Power," *WP*, October 19, 1969.

218 **"Either there's something":** Doel pt. 5.

218 **"Safety is a":** NRC, *Report to the US Atomic Energy Commission.*

219 **"have not resulted":** Quoted in Glenn Seaborg to Frederick Seitz, November 1, 1965 (NAS archives, series "NAS-NRC Executive Office/Records, Earth Sciences: Com on Geologic Aspects of Radioactive Waste Disposal," folder: "Background Info [1967]").

219 **"entirely collusive" and "There has never":** Luther Carter, *Nuclear Imperatives and Public Trust* (Washington, DC: Resources for the Future, 1987), p. 65.

220 **"beyond the requested":** Seaborg to Frank Church, October 17, 1969, in *Congressional Record*—Senate, March 6, 1970, S3139.

220 **"utter disaster":** Quoted in J. Samuel Walker, *Containing the Atom* (Berkeley: University of California Press, 1992), pp. 404.

221 **blow the whistle:** Doel pt. 5.

221 **"I found myself":** Ibid.

221 **"respected world authority":** "Public Fights A-Power," *WP*, October 19, 1969.

221 **"jackpot of energy" and "It has become mandatory":** "Nuclear Power Now?" *Seattle Times*, November 24, 1969.

221 **"We may even":** "Public Fights A-Power," *WP*, October 19, 1969.

221 **"According to AEC":** Dialogue in this and the next paragraph is from Foreman, *Nuclear Power*, p. 239, 254.

222 **"the atomic energy establishment":** Foreman, *Nuclear Power and Public*, p. 124.

222 **"persistently withheld by":** MKH, "Addendum submitted by M. King Hubbert" (AHC, B156 F "Nuclear Power and the Public, October 10–11, 1969 [1]"); excerpt reprinted in US Senate, *Underground Uses of Nuclear Energy*, p. 452.

222 **"I am increasingly troubled":** *Congressional Record*—Senate, March 6, 1970, p. S3139.

223 **"I have only":** MKH, "Telephone Call from Mr. Curtis Wilkie, Assistant to Senator Mondale," memo, March 18, 1970 (AHC, B60 F1).

CHAPTER 29: ENERGY CRISIS

225 **Nixon ignored:** Yergin, *Prize*, p. 589.

225 **"energy crisis":** For example, "Switch May Not Turn On the Lights," UPI in *Times-News* (Hendersonville, NC), May 25, 1970.

226 **"We've given up hoping":** "The Long Cold Winter—II," AP in *Sumter* (SC) *Daily Item*, December 8, 1970.

226 **"We see a potentially":** US House, *Tariff and Trade Proposals, Part 8*, p. 2223.

226 **"domestic production of":** Ibid., p. 2285.

227 **"This is the first":** MKH to Earl Cook, August 4, 1970 (AHC, B127 F "Cook, E. [4]").

227 **"We're somewhere right":** "Man's Conquest of Energy," transcript, September 28, 1970 (AHC, B137 F "Man's Conquest of Energy").

227 **"Were it not for":** National Petroleum Council, *US Energy Outlook*, p. 25.

227 **"The possibility of":** "Fiction & Fact," *OGJ*, May 24, 1971.

228 **"well below our":** M. A. Wright, "US Energy Crisis and What Can Be Done About It," *Ocean Industry*, June 1971.

228 **"The crisis is":** "Fiction & Fact," *OGJ*, May 24, 1971.

228 **"given adequate economic":** National Petroleum Council, *US Energy Outlook*, cover letter.

CHAPTER 30: LIMITS

229 **"It would be deeply":** Henry Jackson to Rogers Morton, July 23, 1971 (AHC, B44 F12).

229 **"Did you hear about":** Quoted in Richard Reeves, *Old Faces of 1976* (New York: Harper & Row, 1976), p. 93.

229 **a hand puppet:** Ibid, p. 56.

230 **"I signed that":** Doel pt. 8.

230 **"I could hardly":** MKH, "File Memorandum (Confidential)," November 5, 1971 (AHC, B170 F "Senate Interior and Insular Affairs—File Memos").

231 **"slipping productive capacity"**: "US Heading for a Close Shave on Crude This Year," *OGJ*, February 14, 1972.

231 **"America has always"**: "ZPG: A New Movement Challenges the US to Stop Growing," *Life*, April 17, 1970.

232 **"The idea of a zero"**: "Is Zero Growth in Our Future?" *Sarasota* (FL) *Herald-Tribune*, February 1, 1970.

232 **"There it was"**: Donella Meadows, *The Global Citizen* (Washington, DC: Island Press, 1991), p. 2. The article was "An End to All This," *Playboy*, July 1971.

232 **"administrative impediments"**: MKH, note on call from Dennis Meadows, January 6, 1972 (AHC, B47 F15).

233 **"physical limits to"**: "Collapse of World Economy Foreseen If Growth Goes On," *LAT*, March 3, 1972.

233 **"It's a simplistic"**: "Mankind Warned of Perils in Growth," *NYT*, February 27, 1972.

233 **"A society released"**: Ibid.

234 **The official launch**: "The Limits to Growth: Hard Sell for a Computer View of Doomsday," *Science* 175, no. 4026 (March 10, 1972), pp. 1088–92, doi:10.1126/science.175.4026.1088.

234 **"Growth is a cop-out"**: Anthony Lewis, "Ecology and Politics: II," *NYT*, March 6, 1972.

234 **"reached some kind"**: MKH, "Comments Prepared for the Secretary of Interior" (AHC, B159 F "Rome, Club of; Limits to Growth").

235 **"Texas in for"**: "Texas in for Year-Round, All-Out Production," AP in *DMN*, March 19, 1972.

235 **"Maybe that will" and "pound of flesh"**: Ibid.

235 **"Old Hubbert was right"**: Deffeyes, *Hubbert's Peak*, pp. 4–5.

CHAPTER 31: THE REPUBLICAN ETHIC

237 **"Things here are buzzing"**: MKH to David Willis, September 6, 1972 (AHC, B56 F1).

237 **"the one petroleum expert"**: Stewart Udall, "The Last Traffic Jam," *Atlantic*, October 1972.

237 **"After all these years"**: MKH to Willis, September 6, 1972, (AHC, B56 F1).

237 **"What on earth"**: *Fortune* magazine advertisement, *NYT*, September 13, 1972.

237 **"It should scare"**: Conversation no. 794-2, October 9, 1972, Nixon Materials, US National Archives and Records Administration, quoted in Hakes, *Declaration of Energy Independence*, p. 21.

238 **"stagflation"**: " 'Stagflation'—A New Word," *Tucson* (AZ) *Daily Citizen*, February 18, 1971.

238 **"Nixonomics"**: One of many examples was "Nixonomic Policies Make Have-Nots Have-Notter," *Raleigh* (Beckley, WV) *Register*, January 19, 1972.

238 **"rapidly increasing dependence"**: Pratt, Becker, and McClenahan, *Voice of Marketplace*, p. 92.

238 **"The era of"**: "Peterson Assures Oil Industry US Will Act on Energy Needs," *NYT*, November 15, 1972.

239 **"They nearly worked"**: MKH to Rubey, August 1, 1973 (Rubey papers, B5 F11).

239 **"Hubbert is one"**: "When Our Resources Run Out," *San Francisco Chronicle*, May 14, 1973.

239 **"the monetary culture"**: MKH, notes for talk, "Social Implications of a Finite World," April 12, 1973 (AHC, B131 F "Energy—analysis of estimates").

239 **"It was insured"**: Ibid. See also Doel pt. 4.

239 **"strenuous and intellectually"**: MKH to Stewart Udall, May 28, 1973 (AHC, B53 F20).

240 **"There is a very"**: MKH to Wallace Pratt, February 24, 1973 (AHC, B50 F13).

240 **James Akins:** Based on "Top Energy Expert: James E. Akins," *NYT*, April 19, 1973; "James Akins; Envoy Foresaw '73 Oil Embargo," *NYT*, July 26, 2010; "James Akins, 83, Dies," *WP*, July 27, 2010; Knowles, *America's Energy Famine*.

240 **"Mr. Akins, you've got"**: Quoted in McQuaig, *It's the Crude*, p. 288. Alternate versions are quoted in Yergin, *Prize* p. 573; Peter Carroll, *It Seemed Like Nothing Happened: America in the 1970s* (New Brunswick, NJ: Rutgers University Press, 1990), p. 120.

240 **"This will not"**: Quoted in Stewart Udall, "The End of an Era," *WP*, January 13, 1974; see also "Decades of Inaction Brought Energy Gap," *NYT*, February 10, 1974.

240 **little experience with oil:** Cooper, *Oil Kings*; Knowles, *America's Energy Famine*, p. 9.

240 **"Nixonese"**: "Top Energy Expert: James E. Akins," *NYT*, April 19, 1973.

241 **"We can now"**: Quoted in Robinson, *Yamani*, p. 76.

241 **"They need us" and "the threat to use"**: Quoted in Akins, "Oil Crisis."

241 **Saudi Arabia . . . was acutely aware:** Knowles, *America's Energy Famine*.

241 **"buyer's cartel"**: Mohammed Ahrari, *OPEC: The Failing Giant* (Lexington: University of Kentucky Press, 1986), pp. 91, 94; see also Akins, "Oil Crisis."

241 **"If it is war"**: "Energy Crisis: Shortages Amid Plenty," *NYT*, April 17, 1973.

242 **"We should not be misled"**: Quotes in this paragraph and the next four are from Nixon, "Special Message to the Congress on Energy Policy," April 18, 1973.

243 **"it means we" and "Oil without a market"**: Nixon, "The President's News Conference," September 5, 1973.

243 **"pitiful"**: Masters to MKH, August 26, 1973 (AHC, B41 F17).

244 **"The study you"**: MKH to Charles Masters, September 7, 1973 (AHC, B41 F17).

245 **"approach an environmental calamity"**: MKH, *US Energy Resources*, p. 194.

246 **"premeditated plan"**: Knowles, *America's Energy Famine*, p. 61.

246 **"Our institutions, our system" and "We're going into"**: "Growth Outpaces Fuel," *San Mateo* (CA) *Times*, October 12, 1973.

CHAPTER 32: 180 DEGREES

249 **"The Saudis are":** Material in this and the next five paragraphs is from James Schlesinger, in Memorandum of Conversation (Kissinger, Schlesinger, Colby, Moorer, Scowcroft), November 3, 1973, 8:47–9:50 a.m., National Security Adviser's Memoranda of Conversation Collection, Gerald R. Ford Presidential Library, http://www.fordlibrarymuseum.gov/Library/document/0314/1552628.pdf. The quotes are from notes by Major General Brent Scrowcroft, deputy assistant to the president for national security affairs, who was in the meeting. Schlesinger later confirmed these plans: "I was prepared to seize Abu Dhabi. . . . It wasn't just bravado. It was clearly intended as a warning" (Robinson, *Yamani*, p. 102). On November 21, 1973, Kissinger warned in a press conference that "if pressures continue unreasonably and indefinitely, then the United States will have to consider what countermeasures it would have to take," and he noted in his memoir: "the United States would not simply passively endure blackmail. At a given point, retaliation was probable. . . . These were not empty threats" (Kissinger, *Years of Upheaval*, p. 880). Also see Cooper, *Oil Kings*, pp. 129–30.

251 **"blackmail":** Kissinger, *Years of Upheaval*, chap. 19.

251 **"We are heading":** Quotes in this paragraph and the next are from Nixon, "Address to the Nation about Policies to Deal with the Energy Shortages," November 7, 1973.

251 **"man killer":** Doel pt. 8.

252 **"Audiences are no longer":** MKH, "Telephone Call from Stewart L. Udall," memo, January 3, 1974 (AHC, B60 F3).

252 **handling of nuclear wastes:** Carroll Wilson, the first general manager of the AEC, concurred with Hubbert's view, writing, "Chemists and chemical engineers were not interested in dealing with waste. It was not glamorous; there were no careers; it was messy; nobody got brownie points for caring about nuclear waste. The Atomic Energy Commission neglected the problem." Wilson, "Nuclear Energy."

252 **"the only source":** "Breeding Power," *Newsweek*, June 14, 1971.

252 **"180 degrees":** "'Convert' Stresses N-Power Hazards," *Denver Post*, October 27, 1974.

252 **"Fifteen years ago":** Ibid.

252 **"this unsteady world" and "hold up New York":** "A Grim View of Nuclear Energy," *San Francisco Examiner*, October 3, 1973.

253 **"It doesn't take":** "Technology Viewed as Key to Energy," *Milwaukee Journal*, December 5, 1973.

253 **"perpetual hazard":** "Solar Power Not Nuclear Called Best," AP in *Victoria* (TX) *Advocate*, January 16, 1974.

253 **"the technological difficulties":** MKH, "Energy Resources of Earth."

253 **"Solar energy dwarfs":** "US Official Calls the Energy Crisis Key History Event," *Buffalo* (NY) *Evening News*, September 14, 1973.

253 **"It turns out"**: "Sunshine Termed Light for the Future," *San Antonio Express*, April 4, 1974.

253 **"I'm happy to say"**: "A New Cultural Shock," *Buffalo* (NY) *Evening News*, September 25, 1973.

253 **"The Saudis are blinking"**: This dialogue is from Memorandum of Conversation (Kissinger, Schlesinger, Colby, Moorer, Scowcroft), November 29, 1973, 1:20–2:38 p.m., National Security Adviser's Memoranda of Conversation Collection, Gerald R. Ford Presidential Library, http://www.fordlibrarymuseum.gov/library/document/0314/1552636.pdf.

254 **"We want Nixon"**: "The New Highway Guerrillas," *Time*, December 17, 1973.

254 **"motorists would follow"**: "Memories of Gas Lines 25 Years Ago," *Morning Call* (Lehigh Valley, PA), October 25, 1998.

254 **"I would say" and "A government policy"**: Quoted in Karen R. Merrill, *The Oil Crisis of 1973–1974: A Brief History with Documents* (New York: St. Martin's, 2007), p. 78.

254 **"triumph of scare"**: "Pipeline Wins," *NYT*, November 14, 1973.

255 **"The Big Car"**: "The Big Car: The End of the Affair," *Time*, December 31, 1973.

255 **"so that we can release"**: "Gov't to Free Reserves When Arab Oil Flows," AP in *Lewiston* (ME) *Daily Sun*, March 14, 1974.

255 **"cultural shock"**: "Growth Outpaces Fuel," *San Mateo* (CA) *Times*, October 12, 1973.

CHAPTER 33: THIS SACRED THING OF GROWTH

256 **"national influence"**: Doel pt. 8.

256 **"inner club"**: Eric Redman, *The Dance of Legislation* (Seattle: University of Washington Press, 2001), p. 80.

256 **"We'd like you"**: This dialogue is from Doel pt. 8.

257 **"prevent or defuse"**: US Senate, *Domestic Supply Information Act*, hearings, p. 2.

257 **"We have never before"**: Ibid.

257 **"That can very definitely"**: Quotes in this and the next paragraph are in ibid., pp. 228–34.

259 **"His estimates turned out"**: Ibid., pp. 259–60.

259 **"uniformly expansive oil outlook"**: Quotes in this paragraph and the next are from Udall, Conconi, and Osterhout, *Energy Balloon*, pp. 7–9.

260 **"Last year, in 1973"**: US House, *National Energy Conservation Policy Act*, hearings, p. 1.

260 **"farsighted scientists, businessmen"**: Quotes in this and the next paragraph are in ibid., pp. 51, 78.

261 **"Events are not"**: Abelson, "Energy Crunch."

261 **"The prospects have"**: US Department of the Interior, "USGS Releases Revised

US Oil and Gas Resource Estimates," news release, March 26, 1974, reprinted in NRC, *Mineral Resources and Environment*, appendix to sec. 2.

261 **"For the first time"**: MKH to Wallace Pratt, May 13, 1974 (AHC, B50 F13).

262 **"where in the world"**: Quotes in this and the next paragraph are from John Moody to Vincent McKelvey, April 8, 1974 (AHC, B155 F "NAS-NRC COMRATE meeting").

262 **"He sent copies"**: Doel pt. 9.

263 **the workshop**: "Second Draft Minutes of Panel III Meeting," July 8, 1974 (NAS archives, series "NAS-NRC Executive Offices," F "COMRATE—General 1974"); Gillette, "Oil and Gas Resources."

263 **"optimistic but not unreasonable"**: William E. Mallory, "Synopsis of Procedure," January 28, 1974, republished in NRC, *Mineral Resources and Environment*, appendix to sec. 2.

263 **"Well, that's one hypothesis"**: Doel pt. 9.

263 **"mulling it over"**: Gillette, "Oil and Gas Resources."

CHAPTER 34: BLUEPRINTS

265 **"this level of production"**: US Federal Energy Administration, *Project Independence Report* (Washington, DC: GPO, 1974), vol. 1, p. 82.

265 **"The oil and gas"**: Ibid., p. 443.

265 **"One of the things we hear"**: "Kissinger on Oil, Food, and Trade," *Business Week*, January 13, 1975, pp. 66–76.

265 **"Actual Strangulation"**: "Actual Strangulation Requires Definition," *Sarasota* (FL) *Journal*, January 14, 1975.

266 **"I would re-affirm"**: "Ford Says He Backs Kissinger on Force if Essential in Mideast," *NYT*, January 13, 1975.

266 **"no one denies"**: Ignotus, "Seizing Arab Oil"; also see Robert Dreyfuss, "The Thirty-Year Itch," *Mother Jones* (March–April 2003).

266 **"could be swiftly"**: Collins and Mark, *Oil Fields as Military Objectives*, p. 16.

266 **"constant security against"**: Ibid., pp. 75–76.

266 **"Anyone who would"**: James Akins, "Now That Our Troops Are in the Oil Fields, Will We Let Go?" *LAT*, September 12, 1990; also see: Dreyfuss, "Thirty-Year Itch."

267 **"kind of a national"**: "Research Study Says 1980 Energy Goal Impossible," AP in *Morning Record* (Meriden, CT), February 12, 1975.

267 **"clearly draws its"**: "A Scramble for Oil That May Not Be There," *Business Week*, March 10, 1975.

267 **"in reality are"**: US Congress, *Adequacy of Oil and Gas Reserves*, hearing, p. 60.

267 **"the impression that"**: Quoted in "Energy Policy Document Termed Too Optimistic," *LAT*, March 19, 1975.

267 **"The realization had" and "very, very jumpy"**: Doel pt. 8.

268 **"Vincible McWelldry"**: This and other quotes this paragraph are from US Geological Survey, *Administrative Unrest*, no. 100,001, April 1, 1975 (AHC B50 F5).

268 **"they didn't know"**: Rose interview by author.

269 **"I got an irate"**: Quotes in this and the next paragraph are from Miller interview by author.

270 **"a zero recovery"**: Magida, "USGS Plays."

270 **"Oil, Gas Estimates"**: "Oil, Gas Estimates Slashed in Survey," UPI in *Beaver County* (PA) *Times*, June 20, 1975.

270 **"Even if the"**: "USGS Re-estimates Reserves," *Geotimes*, August 1975, p. 21.

270 **"a heroic job" and "bloody battle"**: MKH to Ray Walters, August 10, 1977 (AHC, B57 F2).

270 **"the most important report"**: MKH to Michel Grenon, July 29, 1975 (AHC, B44 F5).

271 **"When the United States" and "a prophet without"**: "Expert Doubts US Can Substantially Cut Imports," *NYT*, July 8, 1975.

271 **"acquired a reputation"**: Gillette, "Oil and Gas Resources."

271 **"With modern exploration"**: MKH, "Telephone call from Mr. Bowers, CIA," memo, March 7, 1975 (AHC, B37 F3).

272 **"The OPEC countries"**: "Health Facilities and the Energy Crisis: A Conversation with M. King Hubbert" (Washington, DC: WETA-TV, 1976).

272 **"rather embarrassing"**: MKH to Thomas Nolan, August 2, 1976 (AHC, B48 F18).

272 **"Goodbye King"**: Flyer titled "Goodbye King" (AHC, B78, untitled folder).

CHAPTER 35: BUOYED UP

273 **"central premise"**: Anthony Lewis, "Turning Point," *NYT*, April 4, 1977.

273 **Admiral Hyman Rickover:** Rickover's expectation that US oil production would peak in 1965 is from *Education and Freedom* (New York: Dutton, 1959), p. 28. His wider outlook on energy is in Rickover, "Energy Resources and Our Future," speech to Annual Scientific Assembly of the Minnesota State Medical Association, May 14, 1957, reported on in "The Future of Fossil Fuels," *Christian Science Monitor*, June 5, 1957. On Rickover's influence on Carter's thinking about energy, see Schlesinger interview by author. Rickover and Hubbert did have contact at least once, in 1972, when Rickover missed a talk by Hubbert and asked for a copy to be sent to him.

274 **"Superbrain's Superproblem"**: "Superbrain's Superproblem," *Time*, April 4, 1977.

274 **"as yet no"**: Ibid.

274 **"Americans have never"**: Quotes in this paragraph are from Lewis, "Turning Point."

275 **"It is a little"**: "Nobody Listened in 1956," *Newsday*, April 17, 1977.

275 **"Tonight I want to have":** The quotes from Carter's speech and Hubbert's annotations are from White House, "President Carter's 'Fireside Chat' on Energy," April 18, 1977 (AHC, B37 F13).

276 **"I'm excited over":** MKH to Bretz, May 9, 1977 (Bretz papers, B7 F14).

276 **"healthy economic growth" and "Growth!":** White House, "President Carter's 'Fireside Chat.'"

277 **"Don't let that thing":** Schlesinger interview by author; also see "Good Bye, Dr. McKelvey," *WSJ*, September 16, 1977.

277 **"The CIA report is":** White House, "CIA's International Energy Outlook," April 18, 1977 (AHC, B37 F13).

277 **"Soviet oil production":** US CIA, "International Energy Outlook."

277 **"Sweeping statement with":** Ibid., MKH annotation.

277 **"Seminar attended by":** US embassy Stockholm to US Information Agency, Washington, DC, cable, April 22, 1977 (AHC, B54 F5).

277 **"obviously buoyed up":** US embassy Brussels to US Information Agency, Washington, DC, cable, April 22, 1977 (AHC, B54 F5).

CHAPTER 36: INCOMPREHENSIBLY LARGE NUMBERS

278 **"wanted his own team":** "McKelvey Says Andrus Wanted Own Team in Geological Survey," Gannett in *Sioux Falls* (SD) *Argus-Leader*, September 20, 1977; see also McKelvey to Cecil Andrus, July 11, 1977 (McKelvey papers, B3 F "McKelvey resignation [1 of 3]").

278 **"This is an almost":** Quoted in "Good Bye, Dr. McKelvey," *WSJ*, September 16, 1977.

279 **"the most disgraceful epoch":** MKH to Ray Walters, August 10, 1977 (AHC, B57 F2).

279 **"it would be deplorable":** "When Science Has to Be Political," *WP*, September 13, 1977; see also Philip Abelson, "Leadership of the Geological Survey," *Science* 198, no. 4312 (October 7, 1977), p. 11, doi:10.1126/science.198.4312.11; Nolan to McKelvey, August 27, 1977 (McKelvey papers, B3 F "McKelvey resignation [1 of 3]").

279 **"Apparently the story":** MKH to Bretz, September 17, 1977 (AHC, B57 F2).

280 **"repeatedly expressed his":** "Environmentalists Oust Another Top Professional," *OGJ*, September 9, 1977.

280 **"because of his optimistic":** Jack Kemp, "Dr. Vincent McKelvey, Was He Replaced for Being Too Optimistic about Our Domestic Sources of Energy?" *Congressional Record,* September 14, 1977, pp. E5561–62.

280 **"Russian scientists now":** Ronald Reagan, *Reagan's Path to Victory: The Shaping of Ronald Reagan's Vision: Selected Writings*, ed. Kiron K. Skinner, Annelise Anderson, and Martin Anderson (New York: Free Press, 2004), pp. 206–8.

280 **"While you are trying":** "Good Bye, Dr McKelvey," WSJ, September 16, 1977.

281 **"run out":** "The 'Energy Crisis' Explained," *WSJ*, May 27, 1977.

281 **"1,001 Years of"**: "1,001 Years of Natural Gas," *WSJ*, April 27, 1977.
281 **"20 million years" and "There are plenty"**: "20 Million Years of Energy," *WSJ*, Sep September 14, 1977.
281 **"All that's needed"**: "The Special Interests," *WSJ*, June 17, 1977.
281 **Gallup polls found:** "Gallup Says Carter Convinces Public," AP in *Daytona Beach* (FL) *Morning Journal*, April 25, 1977; and "A Crisis of Credibility," *NYT*, June 12, 1977.

CHAPTER 37: THE PROPHET

282 **"It was not"**: Quoted in "Marion K. Hubbert Dies," *WP*, October 14, 1989.
282 **"Oil Prophet Cited"**: "Oil Prophet Cited: Geologist Saw Crisis in '48," *WP*, November 15, 1977.
282 **"all hell has"**: MKH to Bretz, December 7, 1977 (Bretz papers, B7 F14).
283 **"So accustomed am I"**: MKH to Harold Margulies, November 25, 1977 (AHC, B57 F1).
283 **a title Philip Abelson had bestowed:** Abelson, "Energy Crunch."
283 **"the adequacy of"**: US EIA, *Annual Report to Congress 1977*, vol. 1, p. 38; see also: 42 U.S. Code § 7135(a)(2).
283 **"charged with looking" and "In the past"**: US House, *National Energy Policy Institute Act*, hearings, p. 1.
284 **"As I read"**: Quotes in this paragraph and the next three are from ibid., pp. 21–36.
286 **the EIA tried employing:** US EIA, *Reexamination of Estimation*.
286 **"Hubbert factors"**: US EIA, *Documentation of Volume Three*.
286 **"extremely weak"**: Gass, *Validation and Assessment*, pp. 489–93.
287 **"actual production will probably fall"**: This quote and the Hubbert-inspired graph are from US EIA, *Annual Report to Congress 1978*, vol. 3, *Forecasts*, pp. 13–14.

CHAPTER 38: HUBBERT FACTORS

288 **"Iran is not"**: Quoted in Melvin A. Goodman, *Failure of Intelligence: The Decline and Fall of the CIA* (Lanham, MD: Rowman & Littlefield, 2008), p. 100.
289 **"a new national"**: "'Cheaper Crude or No More Food' Is a Heartland Hit," *NYT*, May 6, 1979.
289 **"The energy crisis"**: "Energy Crisis Still Doubted by Public," AP in *Free Lance-Star* (Fredericksburg, VA), May 4, 1979; see also "Public Convinced Oil Crisis a Fake," AP in *Free Lance-Star*, June 4, 1979.
289 **"magic cures" and "One of the most"**: Quoted in Mattson, *What the Heck*, p. 98.
289 **"I began to ask"**: Quotes in this and the next three paragraphs are from Carter, "Address to the Nation on Energy and National Goals: 'The Malaise Speech,'" July 15, 1979.

290 **"almost a sermon":** Quoted in Mattson, *What the Heck*, p. 159.

290 **"most important appointment":** "The Man Who Offers Pain," *Time*, February 19, 1979.

290 **Carter felt he had little choice:** Carter, *White House Diary*, pp. 341–42.

290 **"The depth of our":** "DOE's Schlesinger Plans Other Press Conferences," UPI in *Florence* (AL) *Times Daily*, July 21, 1979.

291 **Schlesinger's reading of geopolitics:** Schlesinger interview by author.

291 **"I was going to give":** Ibid.

291 **"the greatest economic":** "Exit to Persian Gulf: 'Choke Point' for West's Economy," *NYT*, June 7, 1979.

291 **"Let our position":** Carter, "The State of the Union Address Delivered before a Joint Session of the Congress," January 23, 1980.

292 **"get the hell":** Quoted in Mattson, *What the Heck*, p. 184.

292 **"We learned from" and "When it comes":** "Reagan Gives Out False Oil Figures," *St. Petersburg* (FL) *Times*, March 17, 1980.

CHAPTER 39: HEROIC DREAMS

293 **"as misleading as":** MKH to George Pazik, January 8, 1981 (AHC, B158 F "Pazik").

293 **a mission to extract gold:** Thomas Hager, *The Alchemy of Air* (New York: Harmony Books, 2008), chap. 15.

294 **"the effect may easily":** MKH, "Techniques of Prediction as Applied," pp. 140–41. Hubbert made a similar argument in MKH to A. F. van Everdingen, January 25, 1978 (AHC, B75 F2). For more on the idea of "energy return on investment," see Mason Inman, "The True Cost of Fossil Fuels," *Scientific American* 308, no. 4 (April 2013), doi:10.1038/scientificamerican0413-58.

294 **"That man was":** William Engdahl, "Michel Halbouty Discusses the US Oil Exploration Potential," *Executive Intelligence Review*, May 19, 1981.

294 **"prediction by ambiguous":** MKH, "Techniques of Prediction with Application," p. 3.

294 **"We're going to find":** Engdahl, "Michel Halbouty."

294 **"There is as much oil":** "Production versus Consumption," *Petroleum Review*, October 1980.

295 **"Reagan and I":** Margonelli, *Oil on the Brain*, p. 96.

295 **"We've just got":** "How the US Might Find a Lot More Oil," *NYT*, February 1, 1981.

295 **"one hour after":** "Reagan Pressured to Decontrol Oil Price," *CT*, January 18 1981.

296 **"These United States":** Reagan, "Inaugural Address," January 20, 1981.

296 **"get government off our backs":** "Reagan Theme: A 'Very Personal' Presidency," *Christian Science Monitor*, November 10, 1980.

296 **"It's hard to" and "Our national wealth":** "Growth Day 1981," *NYT*, April 9,

1981, p. B15. On the background of Growth Day, see Kim Phillips-Fein and Julian E. Zelizer, eds., *What's Good for Business* (Oxford: Oxford University Press, 2012), chap. 12.

296 **"demonstrates much more"**: MKH to Ernst Berndt, September 2, 1981 (AHC, B65 F "MIT Energy Laboratory").

297 **"near optimal manner"**: Quotes in this and the next two paragraphs are from MKH, "Two Incompatible Intellectual Systems," handwritten notes for MIT talk (AHC, B131 F "Energy—notes, reprints").

297 **"stupidity"**: Clark, "King Hubbert."

297 **"My departure was"**: Doel pt. 4.

298 **"Hubbert's estimates were"**: C. B. Raleigh, "The Vetlesen Prize, to M. King Hubbert," *Eos* (April 27, 1982).

298 **"Fundamentally I'm an educator"**: Doel pt. 9.

298 **"It's a problem"**: Pazik, "Our Petroleum Predicament."

300 **Daniel Yergin**: "Oil Hindsight," letter, *New Republic*, March 3, 1986; "The Prize," C-SPAN Booknotes, January 27, 1991; "Turning a Prophet: How a Historian Became a Market Guru and Hit the Jackpot," *WP*, April 9, 1998.

300 **"We'd be making"**: "What Happens to OPEC Now with Oil Glut?" Gannett in *Register Star* (Rockford, IL), April 11, 1982.

300 **"We don't have to go"**: "New Oil Crisis Seen by 1990," UPI in *Boston Herald*, July 12, 1982.

300 **"the critical question"**: Yergin, "Awaiting the Next."

300 **"an avalanche of"**: Yergin, "Pangloss on Energy Future."

301 **"The tragedy for Texas"**: "Texas Oil Era Fading," *DMN*, December 19, 1982.

301 **"it's not a good"**: "100 Billion Barrels of Oil Remain Trapped Beneath Texas," *DMN*, December 20, 1982.

301 **"They assume that everything"**: Yergin, "Pangloss on Energy Future."

301 **"wishful thinking"**: Ibid.

301 **"No monetary manipulation"**: MKH to Richard Leo, December 21, 1981 (AHC, B58 F1).

301 **"Buffalo shortage?"**: Ed Stein, cartoon originally published in *Rocky Mountain News*, 1977.

302 **"the only energy analyst"**: William R. Finger and David Nissen, "Oil and Natural Gas Supply Subsystem: A Critique," in Gass, Murphy, and Shaw, *Intermediate Future Forecasting System*, p. 39.

302 **"much scarcer" and "But I am sure"**: MKH, "Telephone call from Jim MacKenzie," memo, December 15, 1982 (AHC, B58 F1).

302 **"I am beginning"**: MKH to Joseph Riva, January 1, 1983 (AHC, B58 F2).

303 **"stymied for lack"**: MKH to Arthur L. Smith, August 2, 1983, ibid.

CHAPTER 40: THINGS WE COULD DO *TOMORROW*

304 **"mentally semi-deranged"**: MKH, Interview by Andrews. Based on Hubbert's description, the drug was likely warfarin.

304 **"strengthened what looked"**: "The Troubling Economics of Oil," *NYT*, October 28, 1984.

305 **"There are no limits"**: Reagan, "Inaugural Address," January 21, 1985.

305 **"the deplorable and"**: MKH, "Telephone call from George Pazik," memo, February 24, 1982 (AHC, B59 F6).

305 **"sabotage"**: "Geophysicist Says Sun Is Energy of Future," *Milwaukee Journal*, March 25, 1983.

305 **"It confirms in devastating"**: MKH to Earl Cook, December 5, 1982 (AHC, B58 F1).

305 **"essentially *hors de combat*"**: MKH to L. F. Ivanhoe, May 8, 1985 (AHC, B58 F4).

305 **"a general complacency" and "2 trillion barrels"**: Chevron, *World Energy Outlook* (June 1985).

306 **"Since production cannot"**: "Exxon Foresees 'Real' Energy Price Up 50% by 2000," *NYT*, December 19, 1980. See also Exxon Company, USA, *Energy Outlook 1980–2000* (December 1979).

306 **"adequate economic incentives"**: National Petroleum Council, *US Energy Outlook*, cover letter.

306 **"persistent propaganda line"**: MKH to Wallace Pratt, January 28, 1958 (AHC, B13 F13).

307 **"the available supplemental"**: Chevron, *World Energy Outlook* (June 1985).

307 **"the final cost" and "the need for"**: "Exxon Abandons Shale Oil Project," *NYT*, May 3, 1982.

307 **aligned with its interests**: Bowden, "Estimating US Crude Oil"; Bowden, "Social Construction of Validity."

308 **"Unless there is"**: Yergin, "Oil Surplus."

308 **"It may be risky"**: "The Saudis Learn to Live on $25 Billion a Year," *NYT*, July 21, 1985.

308 **"wishful thinking"**: Yergin, "Pangloss on Energy Future."

308 **"Please Lord, Give Me"**: "Texas, Oklahoma and Louisiana Make Slow Comeback from '86 Oil Bust," *LAT*, November 18, 1990.

309 **"All that is filed"**: "Overconfidence Could Revive Energy Crunch," *LAT*, January 19, 1986.

309 **"Gasoline is flowing"**: This and quotes in the next two paragraphs are from Gerald Greenwald, "Guarding Against a Third Energy Shock," *NYT*, August 3, 1986.

309 **"From the standpoint"**: "Chevron Chairman Urges Federal Price Floor on Oil," *LAT*, November 12, 1986.

310 **"Ideology always takes"**: "Chevron Price Plea Puts Pressure on US Policy," *NYT*, November 13, 1986.

310 **"so disorganized that"**: MKH to Nell Jessup, April 16, 1986 (SZ).

310 **"I am still" and "Where have all"**: MKH to Robert C. Kendall, June 29, 1986 (AHC, B58 F5).

311 **"would flood 25 percent"**: J. Hansen et al., "Climate Impact of Increasing Atmospheric Carbon Dioxide," *Science* 213, no. 4511 (August 28, 1981), pp. 957–66, doi:10.1126/science.213.4511.957; copy of paper with MKH notes and underlining (AHC, B171 F "Solar Energy & Synthetic Fuels [annotations]").

311 **"We went into"**: MKH, Interview by Andrews.

EPILOGUE: 1990–2015

315 **"M. King Hubbert should"**: "A Cause for Thanksgiving, Part I," *Barron's*, November 24, 2012.

315 **"The belief that"**: Seth M. Kleinman, et al., "Resurging North American Oil Production and the Death of the Peak Oil Hypothesis," Citi (February 15, 2012), quoted in "Citigroup Says Peak Oil Is Dead," *WSJ*, February 17, 2012.

316 **"age of freedom"**: George Bush, "Address to the Nation Announcing the Deployment of United States Armed Forces to Saudi Arabia," August 8, 1990.

316 **"Laid bare . . . American policy"**: Thomas Friedman, "Confrontation in the Gulf," *NYT*, August 12, 1990.

317 **"convince our friends" and "I am going to keep"**: "Bush Backs Price Probe, Death Penalty," AP in *USA Today*, June 21, 2000.

317 **"It has nothing to do"**: "Rumsfeld: It Would Be a Short War," CBS News, November 15, 2002. For more on the Iraq war and oil, see Coll, *Private Empire*, chap. 11; McQuaig, *It's the Crude*, chap. 1; Suskind, *The Price of Loyalty*; Yergin, *Quest*, chap. 7.

317 **"When there is a regime"**: "Bush Economic Aide Says the Cost of Iraq War May Top $100 Billion," *WSJ*, September 16, 2002.

317 **"I am saddened"**: Alan Greenspan, *The Age of Turbulence* (New York: Penguin, 2007), p. 495.

317 **"skulduggery"**: Collins and Mark, *Oil Fields as Military Objectives*, p. 75.

318 **"Are the peakists"**: Quotes in this paragraph and next two from Yergin, "Imagining a $7-a-Gallon Future"; see also Yergin, *Quest*, chap. 11.

319 **"Soaring oil prices"**: Claude Mandil, foreword to IEA, *Resources to Reserves: Oil & Gas Technologies for the Energy Markets of the Future* (Paris: IEA, 2005).

319 **"I will say to him"**: "Bush Asks Saudi King to Open Oil Spigots," ABC News (online), January 16, 2008.

320 **"when the market"**: "Bush Prods Saudi Arabia on Oil Prices," *NYT*, January 16, 2008.

320 **"The economy has"**: "Would a US Recession Lower Oil Prices? Not Necessarily," *NYT*, January 23, 2008.

320 **"We're in a difficult"**: "Energy Chief: Flat Production Behind Oil Prices," AP in *USA Today*, June 7, 2008.

320 **"nothing short of spectacular"**: IEA, *World Energy Outlook 2012*, p. 108.

321 **shale gas:** An early paper on the potential for shale gas is Edward O. Ray, "Devonian Shale Production, Eastern Kentucky Field," in R. F. Meyer, ed., *The Future Supply of Nature-Made Petroleum and Gas* (Oxford, UK: Pergamon Press, 1977).

321 **"though not for":** US Congress, Joint Economic Committee, *Long-Run Economics of Natural Gas*.

321 **"We tried every":** Quoted in Zuckerman, *Frackers*, p. 59.

322 **"shale gale":** Daniel Yergin and Robert Ineson, "America's Natural Gas Revolution," *WSJ*, November 3, 2009.

324 **"never regains its":** IEA, *World Energy Outlook 2010*, p. 48.

324 **"The age of cheap oil":** Inman, "Has the World."

325 **"eerie calm":** "Markets 'on Edge' over 'Eerie Calm' in Oil Prices," CNBC, June 16, 2014.

326 **"Highly leveraged producers":** Bank of International Settlements, "International Banking and Financial Market Developments," *BIS Quarterly Review*, March 2015.

326 **"The question is":** Ashlyn Loder, "The Shale Industry Could Be Swallowed by Its Own Debt," *Bloomberg*, June 18, 2015.

327 **"None of them":** "Greenlight's Einhorn Attacks Frackers, Says Pioneer Burns Cash," *Bloomberg*, May 4, 2015.

327 **"We keep raising":** "US Oil Production Keeps Rising Beyond the Forecasts," *NYT*, January 24, 2014.

329 **the most rigorous shale gas analysis:** As of the autumn of 2015, the University of Texas team was also doing studies of two major tight oil plays, the Bakken and Eagle Ford, but the results hadn't been published yet.

329 **"too optimistic":** Gürcan Gülen et al, "Fayetteville Shale-Production Outlook," *SPE Economics and Management* 7, no. 2 (April 2014).

329 **"bad news" and "We're setting ourselves":** Inman, "Natural Gas: Fracking Fallacy."

330 **"The industry is":** "Big Oil's Latest Fear: A Price Shock after Spending Cuts," *Bloomberg*, April 22, 2015.

330 **BP released its long-term:** BP, *Energy Outlook 2035* (February 2015).

330 **Exxon's 2015 outlook:** ExxonMobil, *The Outlook for Energy 2015: A View to 2040* (December 2014).

331 **$200 billion of projects:** "Oil Prices Up as Report Says Companies Re-evaluating $1.5 Trillion in Projects," *Houston Chronicle*, September, 21, 2015.

331 **another price spike:** "Big Oil's Latest Fear," *Bloomberg*, April 22, 2015.

332 **"There is plenty":** "World Oil Demand to Peak Around 95-110 mbpd—BP CEO," Reuters, February 4, 2010.

332 **"towards self-imposed peak":** "Is the End of the Oil Era Nigh?" *FT*, March 27, 2013.

332 **"We believe that":** "Yesterday's Fuel," *Economist*, August 3, 2013.

332 **"There will be worldwide":** "Cheap Oil and Amazing Energy—The World According to Dan Yergin," Ozy, July 16, 2015.

333 **Americans were driving less:** Edward L. Morse et al., "Energy 2020: Out of America," Citi (November 2014), fig 73.

333 **Americans quickly began driving:** "Driving Is Making a Comeback in the US: Kemp," Reuters, May 29, 2015.

333 **"We just wrapped":** "GM's China Sales Rise 4.4 Percent Halfway Through 2015," *Detroit News*, July 6, 2015.

334 **misunderstood and misrepresented:** Prominent examples include: "The Bottomless Beer Mug," *Economist*, April 28, 2005; "Jeroen van der Veer, Chief Executive of Shell, Answers Back," *Times* (London), May 22, 2008; Peter M. Jackson, "Why the 'Peak Oil' Theory Falls Down," Cambridge Energy Research Associates, 2006; Smil, *Energy at the Crossroads*; and Yergin, *Quest*, chap. 11, an excerpt of which was printed as "There Will Be Oil," *WSJ*, September 17, 2011.

335 **"the only thing standing":** MKH to Gilluly, April 28, 1955 (AHC, B19 F3).

335 **"I like people":** The quotes in this and the next two paragraphs are from MKH to Nell, September 19, 1925 (SZ).

336 **"We're entering into something":** Pazik, "Our Petroleum Predicament."

BIBLIOGRAPHY

Speeches by US presidents are from the American Presidency Project website, unless otherwise noted. I have omitted Internet addresses for online documents since they often change, but I have tried to provide sufficient information to find them. If they have been removed from their original sites, they may be available through sources such as the Internet Archive, at http://archive.org. In titles of sources (as well as quotes in the main text), abbreviations have been modified to remove periods (i.e., "C.I.A." changed to "CIA").

Abelson, Philip. "The Energy Crunch." *Eos*, January 1974, doi:10.1029/EO055i001p00003.

Ackerman, Frederick. "Origin of Technocracy." *NYHT*, December 25, 1932.

Adamson, Martha, and Raymond I. Moore. *Technocracy: Some Questions Answered*. New York: Technocracy, 1934.

Akin, William E. *Technocracy and the American Dream*. Berkeley: University of California Press, 1977.

Akins, James E. "Now That Our Troops Are In the Middle East, Will We Let Go?" *LAT*, September 12, 1990.

———. "The Oil Crisis: This Time the Wolf Is Here." *Foreign Affairs*, April 1973.

Board of Economic Warfare. "Questioning of M. King Hubbert." April 14, 1943, transcript, TIP.

Bowden, Gary. "Estimating US Crude Oil Resources: Organizational Interests, Political Economy, and Historical Change." *Pacific Sociological Review* 25, no. 4 (1982): 419–48, doi:10.2307/1388923.

———. "The Social Construction of Validity in Estimates of US Crude Oil Reserves." *Social Studies of Science* 15, no. 2 (1985): 207–40, doi:10.1177/030631285015002001.

Brinkley, Alan. *The End of Reform: New Deal Liberalism in Recession and War*. New York: Alfred A. Knopf, 1995.

Campbell, Colin, and Jean Laherrère. "The End of Cheap Oil." *Scientific American*, March 1998, doi:10.1038/scientificamerican0398-78.

Carter, Jimmy. *White House Diary*. New York: Picador, 2010.

Chase, Stuart. *Technocracy: An Interpretation*. New York: John Day Co., 1933.

Civil Service Commission. "Report of Special Hearing" (questioning of MKH). September 16, 1942. Copy in TIP.

———. "Inserts of Special Hearing" (revisions to questioning of MKH). September 18, 1942. Copy in TIP.

Clark, Robert Dean. "King Hubbert." *Leading Edge*, February 1983, doi:10.1190/1.1438812.

Coll, Steve. *Private Empire: ExxonMobil and American Power*. New York: Penguin, 2012.

Collins, John M., and Clyde R. Mark. *Oil Fields as Military Objectives: A Feasibility Study*. Congressional Research Service report. Washington, DC: GPO, 1975.

Collins, Robert M. *More: The Politics of Economic Growth in Postwar America*. Oxford: Oxford University Press, 2000.

Cooper, Andrew Scott. *The Oil Kings*. New York: Simon & Schuster, 2011.

Deffeyes, Kenneth S. *Hubbert's Peak: The Impending World Oil Shortage*. Princeton, NJ: Princeton University Press, 2001.

Drew, Lawrence J. *Undiscovered Petroleum and Mineral Resources: Assessment and Controversy*. New York: Plenum Press, 1997.

Ellis, Edward Robb. *A Nation in Torment: The Great American Depression 1929–1939*. New York: Kodansha, 1995.

Elsner, Henry, Jr. *The Technocrats: Prophets of Automation*. Syracuse, NY: Syracuse University Press, 1967.

Foreman, Harry, ed. *Nuclear Power and the Public*. Minneapolis: University of Minnesota Press, 1970.

Gass, Saul I. *Oil and Gas Supply Modeling*. Proceedings of a symposium held at the Department of Commerce, Washington, DC, June 18–20, 1980. Washington, DC: GPO, 1982.

———, ed. *Validation and Assessment Issues on Energy Models*. Proceedings of a workshop held at the National Bureau of Standards, US Department of Commerce, Gaithersburg, MD, January 10-11, 1979. Washington, DC: GPO, 1980.

Gass, Saul I., Frederic H. Murphy and Susan H. Shaw, eds. *Intermediate Future Forecasting System*. Proceedings of a Symposium held at the Department of Energy, Washington, DC, August 19, 1982. Washington, DC: GPO, 1983.

Gillette, Robert. "Oil and Gas Resources: Did USGS Gush Too High?" *Science*, July 12, 1974, doi:10.1126/science.185.4146.127.

Hakes, Jay. *A Declaration of Energy Independence*. New York: John Wiley & Sons, 2008.

Hendricks, Thomas A. *Resources of Oil, Gas, and Natural-Gas Liquids in the United States and the World*. US Geological Survey Circular 522. Washington, DC: GPO, 1966.

Hinton, Diana Davids, and Roger M. Olien. *Oil in Texas: The Gusher Age, 1895–1945*. Austin: University of Texas Press, 2002.

Hubbert, M. King. "Bretz's Baraboo, Wisconsin, Field Course: An Example of Superb

Scientific Pedagogy." Address to the Association of Geology Teachers at Foothill College, March 19, 1966, Bretz papers, B3 F10.

———. "Debunking a Myth." Draft in AHC, B60A F "Schlumberger." Published as "Horizontal Fractures: Debunking a Myth." *Technical Review* 34, no. 3 (October 1986): 4–9.

———. "Degree of Advancement of Petroleum Exploration in the United States." *AAPG Bulletin* 51 (November 1967): 2207–27, doi:10.1306/5d25c269-16c1-11d7-8645000 102c1865d.

———. "Determining the Most Probable." *Technocracy* ser. A, no. 12 (June 1938), pp. 4–10.

———. "The Direction of the Stresses Producing Given Geologic Strains." *The Journal of Geology* 36, no. 1 (January–February 1928): 75–84, doi:10.1086/623480.

———. "Economic Transition and its Human Consequences." *Advanced Management*. July–September 1941.

———. "Energy from Fossil Fuels." *Science* vol. 109, no. 2823 (February 4, 1949): 103–9, doi:10.1126/science.109.2823.103.

———. "Energy Resources." In National Academy of Sciences–National Research Council, *Resources and Man: A Study and Recommendations*. San Francisco: W.H. Freeman, 1969.

———. *Energy Resources: A Report to the Committee on Natural Resources*. Pub. no. 1000-D. Washington, DC: National Academy of Sciences–National Research Council, 1962.

———. "The Energy Resources of the Earth." *Scientific American* 225 (September 1971): 60–70, doi:10.1038/scientificamerican0971-60.

———. "Entrapment of Petroleum Under Hydrodynamic Conditions." *Bulletin of the AAPG* 37 (August 1953): 1954–2026, doi:10.1306/5ceadd61-16bb-11d7-8645000102c1865d.

———. "Geological and geophysical survey of fluorspar areas of Hardin County, Illinois. Part 2. An exploratory study of faults in the Cave In Rock and Rosiclare districts by the earth-resistivity method." US Geological Survey Bulletin 942. 1944.

———. Interview by Steve Andrews, March 5, 1988, Washington, DC. Unpublished transcript provided by Andrews.

———. Interviews by Ronald Doel, January 4, 1989 (referred to herein as Doel pt. 1), January 10 (Doel pt. 2), January 13 (Doel pt. 3), January 17 (Doel pt. 4), January 20 (Doel pt. 5), January 23 (Doel pt. 6), January 27 (Doel pt. 7), February 3 (Doel pt. 8), February 6 (Doel pt. 9). American Institute of Physics, College Park, MD, Niels Bohr Library & Archives, https://www.aip.org/history-programs/niels-bohr-library/oral-histories/5031-1.

———. "Is Being Quantitative Sufficient?" In D. F. Merriam, ed., *The Impact of Quantification on Geology*. Syracuse, NY: Syracuse University Press, 1974.

———. "The Mineral Resources of Texas." Houston: Shell Development Company, 1958.

———. "Nuclear Energy and the Fossil Fuels." Preprint, March 8, 1956. In AHC, B92.

———. "Nuclear Energy and the Fossil Fuels." *Drilling and Production Practice*. American Petroleum Institute, 1957.

———. "Prospecting for Hydrodynamic Traps of Petroleum and Natural Gas: A Historical Review." Houston: Shell Development Company, June 21, 1958. In AHC, B112 F.

———. "Report on Some Aspects of the United Nations Scientific Conference on the Conservation and Utilization of Resources." September 1949. In AHC, B179, F "1949 UN Meeting."

———. "Book Reviews: *The Future Supply of Oil and Gas*." *Science* 128 (July 25, 1958): 196, doi: 10.1126/science.128.3317.194.

———. "Some Facts of Life." *Technocracy* ser. A, no. 5 (December 1935).

———. "Strength of the Earth." *Bulletin of the AAPG* 29 (November 1945): 1630–53.

———. *Structural Geology*. New York: Hafner, 1972.

———. "A Suggestion for the Simplification of Fault Descriptions." *The Journal of Geology* 35, no. 3 (April–May 1927): 264–69, doi:10.1086/623406.

———. "Techniques of Prediction with Application to the Petroleum Industry." Houston: Shell Development Company, pub. no. 204, 1959. In AHC, B137 F "Notes by MKH."

———. "Techniques of Prediction as Applied to the Production of Oil and Gas." In Gass, *Oil and Gas Supply Modeling*, pp. 16–141.

———. *Technocracy Study Course*. 5th ed. New York: Technocracy, Inc., 1940.

———. "The Theory of Ground-Water Motion." *The Journal of Geology* 48, no. 8, part 1 (November–December 1940): 785-944, doi:10.1086/624930.

———. *The Theory of Ground Water Motion and Related Papers*. New York: Hafner, 1969.

———. "Theory of Scale Models as Applied to the Study of Geologic Structures." *Bulletin of the GSA* 48, no. 10 (October 1, 1937): 1459–520, doi:10.1130/GSAB-48-1459.

———. "Two Intellectual Systems: Matter-Energy and the Monetary Culture." Presented at the MIT Energy Laboratory, Cambridge, MA, September 30, 1981. In AHC, B65 F "MIT Energy Laboratory."

———. "The United Nations Scientific Conference on the Conservation and Utilization of Resources." *Geophysics* 15, no. 1 (January1950): 110–11, doi:10.1190/1.1437570.

———. *US Energy Resources: A Review as of 1972*. Background paper prepared for the US Senate Committee on Interior and Insular Affairs, 93rd Cong, 2nd sess. Serial no. 93-40 (92-75). Washington, DC: GPO, 1974.

———. "World Oil and Natural Gas Reserves and Resources." In US Congressional Research Service, *Project Interdependence: US and World Energy Outlook through 1990: A Summary Report*. Committee Print 95-33. Washington, DC: GPO, 1977.

Hubbert, M. King, and Frank A. Melton. "Gravity Anomalies and Petroleum Exploration by the Gravitational Pendulum." *Bulletin of the AAPG* 12 no. 9 (September 1928), doi:10.1306/3d932814-16b1-11d7-8645000102c1865d.

Hubbert, M. King, and William W. Rubey. "Role of Fluid Pressure in Mechanics of Overthrust Faulting: I." *GSA Bulletin* 72, no. 9 (1969), doi:10.1130/0016-7606 (1961)72[1445:ROFPIM]2.0.CO;2.

Hubbert, M. King, and David G. Willis. "Mechanics of Hydraulic Fracturing." *Petroleum Transactions, American Institute of Mining, Metallurgical, and Petroleum Engineers* 210 (1957): 153–68.

Ignotus, Miles (pseudonym). "Seizing Arab Oil." *Harper's*, March 1975.

Inman, Mason. "Has the World Already Passed 'Peak Oil'?" *National Geographic News.* November 9, 2010.

———. "Natural Gas: The Fracking Fallacy." *Nature*, December 3, 2014, doi:10.1038/516028a.

International Energy Agency. *World Energy Outlook 2010*. Paris: OECD, 2010.

———. *World Energy Outlook 2012*. Paris: OECD, 2012.

———. *World Energy Outlook 2014*. Paris: OECD, 2014.

Kaufman, Scott. *Project Plowshare: The Peaceful Use of Nuclear Explosives in Cold War America*. Ithaca, NY: Cornell University Press, 2013.

Kennedy, David. *Freedom from Fear: The American People in Depression and War, 1929–1945*. Oxford: Oxford University Press, 1999.

Kissinger, Henry. *Years of Upheaval*. Boston: Little, Brown, 1982.

Knowles, Ruth Sheldon. *America's Energy Famine*. Norman: University of Oklahoma Press, 1980.

———. *The Greatest Gamblers: The Epic of American Oil Exploration*. 2nd ed. Norman: University of Oklahoma Press, 1978.

Loeb, Harold. *Life in a Technocracy*. 1933; reprint Syracuse, NY: Syracuse University Press, 1996.

Magida, Arthur. "He Predicted the Oil Shortage 19 Years Ago." *National Observer*, May 10, 1975.

———. "USGS Plays a 'Numbers Game' on Remaining Oil, Gas Resources." *National Journal*, September 27, 1975.

Margonelli, Lisa. *Oil on the Brain*. New York: Broadway Books, 2008.

Mattson, Kevin. *What the Heck Are You Up To, Mr. President?* New York: Bloomsbury, 2009.

McKelvey, Vincent E. "Resources, Population Growth, and Level of Living." *Science* 129, no. 3353 (April 3, 1959): 875–81, doi:10.1126/science.129.3353.875.

McKelvey, Vincent E., et al. "Domestic and World Resources of Fossil Fuels, Radioactive Minerals, and Geothermal Energy." US Geological Survey. Unpublished manuscript, November 1961. In MKH B132 F "Domestic and World Resources."

McQuaig, Linda. *It's the Crude, Dude: War, Big Oil, and the Fight for the Planet*, rev. ed. Toronto: Anchor Canada, 2006.

Miller, Betty, et al. *Geological Estimates of Undiscovered Recoverable Oil and Gas Resources in the United States*. US Geological Survey Circular 725 (1975).

National Energy Policy Development Group. *National Energy Policy*. Washington, DC: GPO, 2001.

National Petroleum Council. *US Energy Outlook: An Initial Appraisal 1971–1985* (July 1971), vol. 1.

National Research Council. *The Disposal of Radioactive Waste on Land: Report of the Committee on Waste Disposal of the Division of Earth Sciences*. Washington, DC: National Academy of Sciences, 1957.

———. *Mineral Resources and the Environment*. Washington, DC: National Academy of Sciences, 1975.

———. *Natural Resources: A Summary Report to the President of the United States*. Washington, DC: National Academy of Sciences, 1962.

———. *Report to the US Atomic Energy Commission: By the Committee on Geologic Aspects of Radioactive Waste Disposal*. Washington, DC: National Academy of Sciences, 1966.

———. *Resources and Man: A Study and Recommendations*. San Francisco: W.H. Freeman, 1969.

National Resources Committee. *Population Problems*. Washington, DC: GPO, 1938.

Netschert, Bruce. *The Future Supply of Oil and Gas*. Baltimore: Johns Hopkins University Press, 1958.

Olien, Roger M., and Diana Davids Olien. *Life in the Oil Fields*. Austin: Texas Monthly Press, 1986.

Oreskes, Naomi, and Erik M. Conway. *Merchants of Doubt*. New York: Bloomsbury Press, 2010.

Pazik, George. "Our Petroleum Predicament." *Fishing Facts*, November 1976.

Perkins, Peggy. "Technocracy Leader Picturesque Person." *Oregonian*, January 22, 1933.

Pettijohn, Francis J. *Memoirs of an Unrepentant Field Geologist*. Chicago: University of Chicago Press, 1987.

Philosophical Society of Texas. *Proceedings of the Annual Meeting at Fort Clark, December 9, 10, 11, 1960*. Dallas: Philosophical Society of Texas, 1960.

Pogue, Joseph, and Kenneth Hill. *Future Growth and Financial Requirements of the World Petroleum Industry*. New York: Chase Manhattan Bank, 1956.

Pratt, Joseph A., William H. Becker, and William M. McClenahan. *Voice of the Marketplace: A History of the National Petroleum Council*. College Station: Texas A&M University Press, 2002.

Pratt, Wallace. "The Impact of Atomic Energy on the Petroleum Industry." In *Background Material for the Report of the Panel on the Impact of Peaceful Uses of Atomic Energy*. Washington, DC: GPO, 1956.

Priest, Tyler. *The Offshore Imperative: Shell Oil's Search for Petroleum in Postwar America*. College Station: Texas A&M University Press, 2007.

Raymond, Allen. "Prophet of Technocracy Surrounded by Legends," *NYHT*, December 17, 1932.

———. "Technocracy Becoming Indoor Sport of Slump." *NYHT*, December 18, 1932.

———. "Technocracy New Puzzle of Scientists in Utopia." *NYHT*, December 15, 1932.

———. "Technocracy Translates Value in Terms of Energy." *NYHT*, December 16, 1932.

———. *What is Technocracy?* New York: Whittlesey House, 1933.

Robinson, Jeffrey. *Yamani: The Inside Story.* New York: Atlantic Monthly Press, 1989.

Rubey, William W., and M. King Hubbert. "Role of Fluid Pressure in Mechanics of Overthrust Faulting: II." *Bulletin of the GSA* 70 (1969): 167–206, doi:10.1130/0016-7606(1959)70[167:ROFPIM]2.0.CO;2.

Scott, Howard. "Technology Smashes the Price System." *Harper's*, January 1933.

———. "Technocracy Speaks." *Living Age*, December 1932.

Scott, Howard (pseudonym "Industrial Engineer"). "Political Schemes in Industry." *One Big Union Monthly*, October 1920.

———. "The Scourge of Politics in the Land of Manna." *One Big Union Monthly*, September 1920.

Scott, Howard, et al. *Introduction to Technocracy.* New York: John Day Co., 1933.

Shellenberger, Michael, et al. "Where the Shale Gas Revolution Came From." Breakthrough Institute, May 2012.

Smil, Vaclav. *Energy at the Crossroads.* Cambridge, MA: The MIT Press, 2003.

Soennichsen, John. *Bretz's Flood: The Remarkable Story of a Rebel Geologist and the World's Greatest Flood.* Seattle: Sasquatch Books 2008.

Stobaugh, Robert, and Daniel Yergin. *Energy Future.* New York: Random House, 1979.

Suskind, Ron. *The Price of Loyalty: George W. Bush, the White House, and the Education of Paul O'Neill.* New York: Simon & Schuster, 2004.

Udall, Stewart, Charles Conconi, and David Osterhout. *The Energy Balloon.* New York: McGraw-Hill, 1974.

United Nations. *Proceedings of the United Nations Scientific Conference on the Conservation and Utilization of Resources*, vol. 1. Lake Success, NY: United Nations, 1950.

US Cabinet Task Force on Oil Import Control. *The Oil Import Question.* Washington, DC: GPO, 1970.

US Central Intelligence Agency (CIA). "International Energy Outlook." April 1977. Copy in AHC, B37 F13.

US Congress. Joint Economic Committee. *Adequacy of US Oil and Gas Reserves.* 94th Cong., 1st sess. Hearing, February 25, 1975. Washington, DC: GPO, 1975.

———. Joint Economic Committee. *The Long-Run Economics of Natural Gas.* 108th Cong., 2nd sess. Hearing, October 7, 2004. Washington, DC: GPO, 2004.

———. Joint Committee on Atomic Energy. *Development, Growth, and State of the Atomic Energy Industry.* 88th Cong., 1st sess. Hearings, February 20-21 (pt. 1) and April 2-5 (pt. 2), 1963. Washington, DC: GPO, 1963.

US Energy Information Administration. *Annual Report to Congress*, vol. 1, *1977.* DOE/EIA-0036/1. Washington, DC: GPO, 1978.

———. *Annual Report to Congress 1978*, vol. 3, *Forecasts.* DOE/EIA-0173/3. Washington, DC: GPO, 1978.

———. *Documentation of Volume Three of the 1978 Energy Information Administration Annual Report to Congress.* Prepared by Kilkeary, Scott, and Associates. DOE/EIA/CR-0456. Washington, DC: GPO, 1980.

———. *A Reexamination of the Estimation of Undiscovered Oil Resources in the US*. Technical Memorandum DOE/EIA-01 03/31. Washington, DC: GPO, 1979.

———. *Annual Energy Outlook 2015*. Washington, DC: US DOE, 2015.

US House of Representatives, Committee on Interior and Insular Affairs, Subcommittee on the Environment. *National Energy Conservation Policy Act of 1974*. 93rd Cong., 2nd sess. Hearings, June–July 1974. Washington, DC: GPO, 1974.

———. Committee on Interstate and Foreign Commerce, Subcommittee on Energy and Power, *Energy Conservation and Oil Policy*. 94th Cong., 1st sess. Hearings. Washington, DC: GPO, 1975.

———. Committee on Interstate and Foreign Commerce, Subcommittee on Energy and Power. *National Energy Policy Institute Act*. 95th Cong., 2nd sess. Hearings, April 11–12, 1978. Washington, DC: GPO, 1978.

———. Committee on Ways and Means. *Tariff and Trade Proposals, Part 8*. 91st Cong., 2nd sess. Hearings, May–June 1970. Washington, DC: GPO, 1970.

US Office of Technology Assessment. *US Natural Gas Availability: Conventional Gas Supply Through the Year 2000: A Technical Memorandum*. Washington, DC: GPO,1983.

US Senate. Committee on Commerce and Committee on Government Operations. *Domestic Supply Information Act*. 93rd Cong., 2nd sess. Hearings, Washington, DC: GPO, 1974.

———. Committee on Energy and Natural Resources. *Energy, An Uncertain Future: An Analysis of US and World Energy Projections through 1990*, pub. 95–157, prepared by Herman T. Franssen. 95th Cong., 2nd sess. Washington, DC: GPO, 1978.

———. Committee on Interior and Insular Affairs. *Report of the National Fuels and Energy Study Group on an Assessment of Available Information on Energy in the United States*. 87th Cong., 2nd sess. Washington, DC: GPO, 1962.

———. Committee on Public Works, Subcommittee on Air and Water Pollution, *Underground Uses of Nuclear Energy*. 91st Cong., 1st sess. Hearings, November 1969. Washington, DC: GPO, 1969.

Walsh, Richard. "The Engineers' Revolution." *Judge*, May 28, 1932.

Wilson, Carroll. "Nuclear Energy: What Went Wrong." *Bulletin of the Atomic Scientists* 35, no. 6 (June 1979).

Wilson, James E. "Hubbert's Curve 'Piqued' Interest." *AAPG Explorer*, January 2008.

———. "Intellectual Battles Were King's Feast." *AAPG Explorer*, January 2008.

———. "King Hubbert: An Irascible Iconoclast." *AAPG Explorer*, December 2007.

Wright, Myron. "US Energy Crisis and What Can Be Done About It." *Ocean Industry*, June 1971.

Yergin, Daniel. "Awaiting the Next Oil Crisis." *NYT*, July 11, 1982.

———. "Ensuring Energy Security." *Foreign Affairs*, March–April 2006.

———. "The Globalization of Energy Demand." CNBC, June 3, 2013.

———. "Imagining a $7-a-Gallon Future." *New York Times*, April 4, 2004.

———. "It's Not the End of the Oil Age." *WP*, July 31, 2005.

———. "An Oil Surplus—For Now." *NYT*, July 8, 1985.

———. "Pangloss on the Energy Future: Wishful Thinking." *NYT*, November 9, 1982.

———. *The Prize: The Epic Quest for Oil, Money and Power*. New York: Simon & Schuster, 1991.

———. *The Quest: Energy, Security, and the Remaking of the Modern World*. New York: Penguin Press, 2011.

———. "Stepping on the Gas." *WSJ*, April 2, 2011.

Zapp, A. D. "Future Petroleum Producing Capacity of the United States." US Geological Survey, Bulletin 1142-H (1962).

———. "World Petroleum Resources." In McKelvey et al., "Domestic and World Resources." USGS, November 1961, unpublished.

Zuckerman, Gregory. *The Frackers: The Outrageous Inside Story of the New Energy Revolution*. New York: Portfolio Penguin, 2013.

RECOMMENDED READING

Aleklett, Kjell, with Michael Lardelli. *Peeking at Peak Oil*. New York: Springer, 2012.

Campbell, Colin, ed. *Peak Oil Personalities*. Skibbereen, Ireland: Inspire Books, 2011.

Deffeyes, Kenneth. *Hubbert's Peak: The Impending World Oil Shortage*. Princeton University Press, 2001.

Gold, Russell. *The Boom: How Fracking Ignited the American Energy Revolution and Changed the World*. New York: Simon & Schuster, 2014.

Galbraith, James K. *The End of Normal: The Great Crisis and the Future of Growth*. New York: Simon & Schuster, 2014.

Leggett, Jeremy. *The Energy of Nations: Risk Blindness and the Road to Renaissance*. London: Routledge, 2013.

Roberts, Paul. *The End of Oil: On the Edge of a Perilous New World*. London: Bloomsbury, 2004.

Strahan, David. *The Last Oil Shock: A Survival Guide to the Imminent Extinction of Petroleum Man*. London: John Murray, 2007.

Zuckerman, Gregory. *The Frackers: The Outrageous Inside Story of the New Energy Revolution*. New York: Portfolio Penguin, 2013.

ACKNOWLEDGMENTS

Thanks to Rick Broadhead, my agent, for taking a chance on a new author with a rough proposal. Thanks for Brendan Curry, my editor, for taking an interest when others didn't. Thanks to Brendan and to his assistant Sophie Duvernoy for their perceptive comments and thorough edits that undoubtedly made this book far better.

Thanks to Steve Andrews, Chris Kuykendall, and Jason Brenno for early help in locating key documents from Hubbert's archives and for sharing their thoughts on oil. Thanks to Chris Nelder for reaching out and talking things through over (too many) beers. For early votes of confidence, thanks to Bill McKibben, Christopher Shaw, Matt Wheeland, and the whole Posse.

As the book took shape, it had many readers of various drafts. Thanks to Gwyn Kirk, Chris Kuykendall (again), David Biello, Jennie Dusheck, Alex Witze, David Lindley, Suzie Zupan, Elizabeth Svoboda, and George Johnson. Thanks also to those researchers who reviewed portions of the manuscript to check it for accuracy. (Any mistakes, of course, are my fault.)

Thanks to the American Institute of Physics for publishing Hubbert's long oral history—doggedly gathered by historian Ronald Doel—and for permitting me to quote from it. Thanks to the American Heritage Center and the Roosevelt Institute for research grants that funded travel to their sites. Thanks to the librarians and archivists at the various sites I visited, who made my stays pleasant and helped me find more than I would have on my own. At the American Heritage Center, thanks to all the staff, especially John Waggoner and Ginny Kilander. At the Franklin D. Roosevelt Presidential Library, special thanks to Kirsten Carter and Virginia Lewick. At the National Academy of Sciences archives, Janice Goldblum. At the Manuscript Reading Room of the Library of Congress, thanks to all the staff who worked efficiently and cheerfully despite large numbers of visitors. Thanks to all those who took the time for long interviews—especially Dennis Meadows, for treating me to a Cosmos Club dinner and loaning the requisite tie.

Thanks to friends who put me up during my travels: M.J. and Michael, Nick and

Sasha, Margaret, and Courtney. Thanks to my colleagues at Near Zero for understanding when I needed to disappear for long stretches.

Thanks to my close family—Mom and Ken, Meralee and Dean—for always being there for me.

Above all thanks to Sarah Bird—my wife, my best friend, and my debating partner—for her unflagging and cheerful support.

CREDITS AND PERMISSIONS FOR FIGURES

Page 58 **Growth curves:** Based on figure in MKH, *Technocracy Study Course*, p. 101.

Page 101 **The strength of rock:** Illustration of Texas being lifted by a crane from Hubbert, M. King, "Strength of the Earth," *Bulletin of the AAPG*, ©1945, reprinted by permission of AAPG, whose permission is required for further use.

Page 119 **World oil production:** Based on figure in United Nations, *Proceedings of Scientific Conference*, p. 101.

Page 143 **US oil production:** Based on MKH, "Nuclear Energy and the Fossil Fuels," *Drilling and Production Practice*, fig. 21.

Page 158 **Hubbert's three-curve method:** Based on MKH, "Methods of Predicting."

Page 162 **The beer-can experiment:** Illustration from Hubbert, M. King, and William W. Rubey, "Role of Fluid Pressure in Mechanics of Overthrust Faulting: I," *Bulletin of the GSA*, 1969.

Page 190 **Discoveries of large oil fields:** Based on MKH, *Energy Resources: Report to Committee*, fig. 37.

Page 211 **Success rate of oil exploration:** Based on MKH, "Degree of Advancement," fig. 15.

Page 287 **World oil production forecasts:** From US EIA, *Annual Report to Congress 1978*, vol. 3, *Forecasts*, p. 13.

Page 323 **US natural gas production forecasts:** Based on MKH, *Energy Resources: Report to Committee*, fig. 47, and US EIA, *Annual Energy Outlook 2015*, table A14.

Page 326 **US oil production forecasts:** Based on MKH, "Nuclear Energy and the Fossil Fuels," Drilling and Production Practice, fig. 21, and US EIA, *Annual Energy Outlook 2015*, table A14.

Page 331 **World oil production forecasts:** Based on MKH, "World Oil and Natural Gas Reserves," fig. XIX-3, and IEA, *World Energy Outlook 2014*, table 3.6.

INDEX

Note: Page numbers in *italics* refer to charts and graphs.